T0318820

Stochastic Calculus for Quantitative Finance

Optimization in Insurance and Finance Set

coordinated by
Nikolaos Limnios and Yuliya Mishura

Stochastic Calculus for Quantitative Finance

Alexander A. Gushchin

ELSEVIER

First published 2015 in Great Britain and the United States by ISTE Press Ltd and Elsevier Ltd

ISTE Press Ltd
27-37 St George's Road
London SW19 4EU
UK

www.iste.co.uk

Elsevier Ltd
The Boulevard, Langford Lane
Kidlington, Oxford, OX5 1GB
UK

www.elsevier.com

Notices

Knowledge and best practice in this field are constantly changing. As new research and experience broaden our understanding, changes in research methods, professional practices, or medical treatment may become necessary.

Practitioners and researchers must always rely on their own experience and knowledge in evaluating and using any information, methods, compounds, or experiments described herein. In using such information or methods they should be mindful of their own safety and the safety of others, including parties for whom they have a professional responsibility.

To the fullest extent of the law, neither the Publisher nor the authors, contributors, or editors, assume any liability for any injury and/or damage to persons or property as a matter of products liability, negligence or otherwise, or from any use or operation of any methods, products, instructions, or ideas contained in the material herein.

For information on all our publications visit our website at http://store.elsevier.com/

British Library Cataloguing-in-Publication Data
A CIP record for this book is available from the British Library
Library of Congress Cataloging in Publication Data
A catalog record for this book is available from the Library of Congress
ISBN 978-1-78548-034-8

Printed and bound in the UK and US

Contents

Preface

The arbitrage theory for general models of financial markets in continuous time is based on the heavy use of the theory of martingales and stochastic integration (see the monograph by Delbaen and Schchermayer [DEL 06]). Our book gives an exposition of the foundations of modern theory of stochastic integration (with respect to semimartingales. It follows traditions of the Strasbourg School of Stochastic Processes. In particular, the exposition is inspired by the monograph by Dellacherie [DEL 72]) in Chapter 1 and by the course by Meyer [MEY 76] in Chapters 2 and 3. In Chapter 1, the so-called general theory of stochastic processes is developed. The second chapter is devoted to detailed study of local martingales and processes with finite variation. The theory of stochastic integration with respect to semimartingales is a subject of Chapter 3. We do not consider vector stochastic integrals, for which we refer to Shiryaev and Cherny [SHI 02]. The last section is devoted to σ-martingales and the Ansel–Stricker theorem. Some results are given without proofs. These include the section theorem, classical Doob's theorems on martingales, the Burkholder–Davis–Gundy inequality and Itô's formula.

Our method of presentation may be considered as old-fashioned, compared to, for example, the monograph by Protter [PRO 05], which begins with an introduction of the notion of a semimartingale; in our book, semimartingales appear only in the final chapter. However, the author's experience based on the graduate courses taught at the Department of Mechanics and Mathematics of Moscow State University, indicates that our approach has some advantages.

The text is intended for a reader with a knowledge of measure-theoretic probability and discrete-time martingales. Some information on less standard topics (theorems on monotone classes, uniform integrability, conditional expectation for nonintegrable random variables and functions of bounded variation) can be found in the Appendix. The basic idea, which the author pursued when writing this book, was to provide an affordable and detailed presentation of the foundations of the theory of stochastic integration, which the reader needs to know before reading more advanced literature on the subject, such as Jacod [JAC 79], Jacod and Shiryaev [JAC 03], Liptser and Shiryayev [LIP 89], or a literature dealing with applications, such as Delbaen and Schchermayer [DEL 06].

The text is accompanied by more than a hundred exercises. Almost all of them are simple or are supplied with hints. Many exercises extend the text and are used later.

The work on this book was partially supported by the International Laboratory of Quantitative Finance, National Research University Higher School of Economics and Russian Federation Government (grant no. 14.A12.31.0007). I wish to express my sincere thanks to Tatiana Belkina for a significant and invaluable assistance in preparing the manuscript.

Alexander GUSHCHIN
Moscow, May 2015

Basic Notation

The symbol □ indicates the end of the proof.

The symbol := means "put by definition".

$\mathbb{R} = (-\infty, +\infty) =$ the set of real numbers, $\mathbb{R}_+ = [0, +\infty)$.

$\mathbb{R}^d = d$-dimensional Euclidean space.

$\mathbb{Q} =$ the set of rational numbers, $\mathbb{Q}_+ = \mathbb{Q} \cap \mathbb{R}_+$.

$\mathbb{N} = \{1, 2, 3, \dots\} =$ the set of natural numbers.

$a \vee b = \max\{a, b\}$, $a \wedge b = \min\{a, b\}$ for $a, b \in \mathbb{R}$.

$a^+ = a \vee 0$, $a^- = (-a) \vee 0$ for $a \in \mathbb{R}$.

$\lim\limits_{s \uparrow\uparrow t} = \lim\limits_{s \to t,\, s < t}$.

$\mathbb{1}_B =$ the indicator function of the set B.

$\mathsf{E} =$ expectation.

$\mathsf{E}(\cdot | \mathscr{G}) =$ conditional expectation with respect to the σ-algebrs $\mathscr{G}$.

$\mathscr{F} \vee \mathscr{G} = \sigma(\mathscr{F} \cup \mathscr{G}) =$ the smallest σ-algebra containing the σ-algebras $\mathscr{F}$ and $\mathscr{G}$.

$\bigvee_{\alpha \in A} \mathscr{F}_\alpha = \sigma\left(\bigcup_{\alpha \in A} \mathscr{F}_\alpha\right) =$ the smallest σ-algebra containing the σ-algebras $\mathscr{F}_\alpha, \alpha \in A$.

List of Statements

EXAMPLES

EXERCISES

REMARKS

PROPOSITIONS

THEOREM

LEMMAS

COROLLARIES

General Theory of Stochastic Processes

1.1. Stochastic basis and stochastic processes

To describe the dynamics of random phenomena in time, the notion of a stochastic basis is used in stochastic calculus.

DEFINITION 1.1.– A *filtration* on a measurable space $(\Omega, \mathscr{F})$ is a nondecreasing family $\mathbb{F} = (\mathscr{F}_t)_{t \in \mathbb{R}_+}$ of sub-σ-algebras of the σ-algebra $\mathscr{F}$: $\mathscr{F}_s \subseteq \mathscr{F}_t \subseteq \mathscr{F}$ for all $s < t$, $s, t \in \mathbb{R}_+$. A *stochastic basis* is a probability space equipped with a filtration, i.e. a stochastic basis is a quadruplet $\mathbb{B} = (\Omega, \mathscr{F}, \mathbb{F}, \mathsf{P})$, where $(\Omega, \mathscr{F}, \mathsf{P})$ is a probability space and $\mathbb{F}$ is a filtration on $(\Omega, \mathscr{F})$.

If a discrete filtration is given, i.e. a nondecreasing family of sub-σ-algebras $\mathscr{F}_0 \subseteq \mathscr{F}_1 \subseteq \cdots \subseteq \mathscr{F}_n \subseteq \cdots \subseteq \mathscr{F}$, then it can be extended to a filtration in the sense of definition 1.1 in a natural way. Namely, put $\mathscr{F}_t := \mathscr{F}_{[t]}$, where $[\cdot]$ is the integer part of a number.

If $\mathbb{F} = (\mathscr{F}_t)_{t \in \mathbb{R}_+}$ is a filtration, then define

$$\mathscr{F}_{t+} := \bigcap_{s:\, s > t} \mathscr{F}_s, \quad t \in \mathbb{R}_+,$$

$$\mathscr{F}_{t-} := \bigvee_{s:\, s < t} \mathscr{F}_s, \quad t \in (0, \infty].$$

Put also $\mathscr{F}_{0-} := \mathscr{F}_0$, $\mathscr{F}_\infty := \mathscr{F}_{\infty-}$.

$\mathscr{F}_t$ (respectively $\mathscr{F}_{t-}$) is usually interpreted as the class of events occurring before or at time t (respectively strictly before t).

DEFINITION 1.2.– A stochastic basis $\mathbb{B} = (\Omega, \mathscr{F}, \mathbb{F}, \mathsf{P})$ is called *right-continuous* if the filtration $\mathbb{F} = (\mathscr{F}_t)_{t \in \mathbb{R}_+}$ is right-continuous, i.e. $\mathscr{F}_t = \mathscr{F}_{t+}$ for every $t \in \mathbb{R}_+$. A stochastic basis $\mathbb{B} = (\Omega, \mathscr{F}, \mathbb{F}, \mathsf{P})$ is called *complete* if the σ-algebra $\mathscr{F}$ is complete (relative to P) and the σ-algebra $\mathscr{F}_0$ contains all sets of measure 0 from $\mathscr{F}$. We say that a stochastic basis $\mathbb{B} = (\Omega, \mathscr{F}, \mathbb{F}, \mathsf{P})$ satisfies the *usual conditions* if it is right-continuous and complete.

Recall that a σ-algebra $\mathscr{F}$ is complete relative to P if $A \in \mathscr{F}$, $\mathsf{P}(A) = 0$, $B \subseteq A$ imply $B \in \mathscr{F}$.

The condition that a stochastic basis is right-continuous is essentially necessary for the development of the theory. The completeness property of a stochastic basis is much less significant. Statements that are proved for a complete stochastic basis either remain valid without the completeness or need a little correction. On the other hand, the completeness property is very undesirable, especially in the statistics of stochastic processes, as it is not preserved under a change of measure to a non-equivalent one. Nevertheless, in this book, we will always assume, unless otherwise stated, that a stochastic basis satisfies the usual conditions.

EXAMPLE 1.1.– Let $\Omega = \mathbb{D}(\mathbb{R})$ be the space of functions $\omega \colon \mathbb{R}_+ \rightsquigarrow \mathbb{R}$ which are right-continuous at every $t \in \mathbb{R}_+$ and have finite left-hand limits at every $t \in]0, \infty[$, or $\Omega = \mathbb{C}(\mathbb{R})$ is the space of continuous functions $\omega \colon \mathbb{R}_+ \rightsquigarrow \mathbb{R}$. Define mappings $X_t \colon \Omega \to \mathbb{R}$, $t \in \mathbb{R}_+$, by $X_t(\omega) = \omega(t)$. Define the σ-algebra $\mathscr{F}_t^0$, $t \in \mathbb{R}_+$, on Ω as the smallest σ-algebra with respect to which all mappings X_s, $s \leqslant t$, are measurable. Finally, put $\mathscr{F} = \mathscr{F}_\infty^0$. It is obvious that $\mathbb{F}^0 = (\mathscr{F}_t^0)_{t \in \mathbb{R}_+}$ is a filtration on $(\Omega, \mathscr{F})$. $X = (X_t)_{t \in \mathbb{R}_+}$ is said to be the *canonical* stochastic process on Ω.

EXERCISE 1.1.– Show that the filtration $\mathbb{F}^0$ in example 1.1 is not right-continuous.

If a stochastic basis does not satisfy the usual conditions, then there is a "minimal" extension of it to a stochastic basis satisfying the usual conditions. The corresponding construction is given in the following exercises.

EXERCISE 1.2.– Let $\mathbb{B} = (\Omega, \mathscr{F}, \mathbb{F} = (\mathscr{F}_t)_{t \in \mathbb{R}_+}, \mathsf{P})$ be a stochastic basis. Put $\mathbb{F}^+ = (\mathscr{F}_{t+})_{t \in \mathbb{R}_+}$. Show that $\mathbb{B}^+ := (\Omega, \mathscr{F}, \mathbb{F}^+, \mathsf{P})$ is a right-continuous stochastic basis.

EXERCISE 1.3.– Let $\mathbb{B} = (\Omega, \mathscr{F}, \mathbb{F} = (\mathscr{F}_t)_{t \in \mathbb{R}_+}, \mathsf{P})$ be a stochastic basis. Put $\mathscr{N} := \{B \subseteq \Omega \colon B \subseteq A \text{ for some } A \in \mathscr{F} \text{ with } \mathsf{P}(A) = 0\}$, $\mathscr{F}^\mathsf{P} := \{A \triangle B \colon A \in \mathscr{F}, B \in \mathscr{N}\}$, $\mathscr{F}_t^\mathsf{P} := \{A \triangle B \colon A \in \mathscr{F}_t, B \in \mathscr{N}\}$, $t \in \mathbb{R}_+$. Show that $\mathscr{F}^\mathsf{P}$ and $\mathscr{F}_t^\mathsf{P}$ are σ-algebras. For $C \in \mathscr{F}^\mathsf{P}$, put $\overline{\mathsf{P}}(C) := \mathsf{P}(A)$ if C has a form $C = A \triangle B$, $A \in \mathscr{F}$, $B \in \mathscr{N}$. Show that $\overline{\mathsf{P}}$ is well defined. Show that $\overline{\mathsf{P}}$ is a probability on $(\Omega, \mathscr{F}^\mathsf{P})$, whose restriction onto $\mathscr{F}$ is P. Show that $\mathbb{B}^\mathsf{P} := (\Omega, \mathscr{F}^\mathsf{P}, \mathbb{F}^\mathsf{P}, \overline{\mathsf{P}})$, where $\mathbb{F}^\mathsf{P} := (\mathscr{F}_t^\mathsf{P})_{t \in \mathbb{R}_+}$, is a complete stochastic basis.

EXERCISE 1.4.– Let $\mathbb{B} = (\Omega, \mathscr{F}, \mathbb{F} = (\mathscr{F}_t)_{t \in \mathbb{R}_+}, \mathrm{P})$ be a stochastic basis. Show that $(\mathbb{B}^+)^\mathrm{P} = (\mathbb{B}^\mathrm{P})^+$.

REMARK 1.1.– It often happens that the stochastic basis $\mathbb{B}^\mathrm{P}$ is right-continuous. For instance, in example 1.1, this is the case if the measure P on $(\Omega, \mathscr{F})$ is such that the canonical process X is a Lévy process (in particular, if X is a Wiener or a Poisson process).

Let $\mathbb{B} = (\Omega, \mathscr{F}, \mathbb{F} = (\mathscr{F}_t)_{t \in \mathbb{R}_+}, \mathrm{P})$ be a stochastic basis satisfying the usual conditions. A stochastic process with values in a measurable space $(E, \mathscr{E})$ is a family $X = (X_t)_{t \in \mathbb{R}_+}$ of measurable mappings X_t from $(\Omega, \mathscr{F})$ to $(E, \mathscr{E})$. We usually indicate the range E of values of a stochastic process, assuming implicitly that a topology is given on E and $\mathscr{E}$ is the σ-algebra $\mathscr{B}(E)$ of Borel sets, i.e. the σ-algebra generated by open sets. If $E = \mathbb{R}^d$, then it is assumed that the topology corresponds to the Euclidean metric. The extended real line $[-\infty, +\infty]$ is assumed to be equipped with the natural topology as well. We do not usually indicate the range of values of a stochastic process, which means that it takes real values, i.e. $E = \mathbb{R}$. It is obvious that a stochastic process X with values in $\mathbb{R}^d$ is a set $(X^1, \ldots, X^d)$ of d real-valued stochastic processes.

Essentially, a stochastic process X is a function of two variables: ω and t. To emphasize this fact, the value of X at outcome ω and at time t will be denoted not only by $X_t(\omega)$ but also by $X(\omega, t)$. In what follows, measurability properties of stochastic processes as functions of two variables, i.e. mappings from the space $\Omega \times \mathbb{R}_+$ equipped with some σ-algebra, to E, play an important role.

A set $B \subseteq \Omega \times \mathbb{R}_+$ is said to be a *random set* if $\mathbb{1}_B$ is a stochastic process.

The *trajectory* of a stochastic process X, corresponding to outcome ω, is the mapping $t \rightsquigarrow X(\omega, t)$ from $\mathbb{R}_+$ to E.

A process X with values in a topological space E is called continuous (or right-continuous, or left-continuous, or right-continuous with left-hand limits) if *all* its trajectories are continuous (respectively right-continuous, or left-continuous, or right-continuous and have left-hand limits for every $t > 0$). To avoid misunderstanding, we emphasize that left-hand limits must belong to E. Thus, if $E = \mathbb{R}$, a right-continuous process with left-hand limits must have a *finite* limit from the left for every $t > 0$. For a right-continuous process with left-hand limits, we use the French abbreviation càdlàg (*continue à droite avec des limites à gauche*, i.e. right-continuous with left-hand limits).

Linear operations on stochastic processes with values in a linear space E are understood pointwise.

If a process X with values in a linear topological space E is càdlàg, then we define two new stochastic processes $X_- = (X_{t-})_{t \in \mathbb{R}_+}$ and $\Delta X = (\Delta X_t)_{t \in \mathbb{R}_+}$ with values in E by

$$X_{0-} = X_0, \qquad X_{t-} = \lim_{s \uparrow\uparrow t} X_s, \quad t > 0,$$

$$\Delta X = X - X_-.$$

Note that $\Delta X_0 = 0$, according to this definition.

Let X be a real-valued right-continuous stochastic process. Assume that $\lim_{t \to \infty} X_t(\omega)$ exists (respectively exists and is finite) for almost all ω. In this case, X_∞ is understood as any mapping from Ω to $[-\infty, +\infty]$ (respectively to $\mathbb{R}$) such that

$$X_\infty(\omega) = \lim_{t \to \infty} X_t(\omega) \quad \text{P-a.s.}$$

It follows from the completeness of the stochastic basis that any such mapping X_∞ is an $\mathscr{F}_\infty$-measurable random variable. We also put $X_{\infty-} := X_\infty$ under the same assumptions.

EXERCISE 1.5.– Prove the assertion concerning the measurability of X_∞ .

DEFINITION 1.3.– A stochastic process Y is called a *modification* of a stochastic process X, if $\mathsf{P}(Y_t \neq X_t) = 0$ for every $t \in \mathbb{R}_+$.

DEFINITION 1.4.– Stochastic processes X and Y are called *indistinguishable* if

$$\mathsf{P}\{\omega: \text{ there exists } t \in \mathbb{R}_+ \text{ such that } X_t(\omega) \neq Y_t(\omega)\} = 0.$$

We will also speak in this case that Y (respectively X) is a *version* of X (respectively Y).

A random set is called *evanescent*, if its indicator is indistinguishable with the process which is identically zero.

Sometimes in the literature, the term "version" is used as a synonym of the term "modification".

It is assumed in the definition of indistinguishability that the corresponding set belongs to the σ-algebra $\mathscr{F}$. Let us note that, for measurable processes X and Y (see definition 1.8 in section 1.3), the completeness of the σ-algebra $\mathscr{F}$ guarantees that this set is measurable, see theorem 1.7.

It is clear that if two processes are indistinguishable, then they are modifications of each other. The converse is generally not true.

EXAMPLE 1.2.– Let $\Omega = [0, 1]$, $\mathscr{F}$ be the σ-algebra of Lebesgue measurable sets, and let P be the Lebesgue measure on $(\Omega, \mathscr{F})$. Put $X_t(\omega) = 0$ for all ω and t,

$$Y_t(\omega) = \begin{cases} 1, & \text{if } t = \omega, \\ 0, & \text{if } t \neq \omega. \end{cases}$$

Then X and Y are modifications of each other, but they are not indistinguishable.

PROPOSITION 1.1.– If X and Y are right-continuous (or left-continuous), and Y is a modification of X, then X and Y are indistinguishable.

PROOF.– For every rational $r \in \mathbb{Q}_+$, put $N_r = \{\omega \colon X_r(\omega) \neq Y_r(\omega)\}$, and let $N = \bigcup_{r \in \mathbb{Q}_+} N_r$. Then $\mathsf{P}(N) = 0$ because Y is a modification of X. However, for $\omega \in \Omega \setminus N$, trajectories of $X.(\omega)$ and $Y.(\omega)$ coincide in all rational points and, hence, in all points due to the assumption of one-sided continuity. □

In this book, we will often encounter a situation where some stochastic processes are constructed from other stochastic processes. Often the result of such a construction is not a concrete stochastic process but any one in a class of indistinguishable processes. The situation here is similar, say, to that occurs when taking the conditional expectation, which is defined as any random variable from a fixed equivalence class of random variables that coincide almost surely. Uniqueness here could be achieved by considering the equivalence classes of indistinguishable stochastic processes. However, this approach has its drawbacks, and its use, in our opinion, would lead to a complication of terminology and notation. However, the reader should keep in mind this circumstance and understand that, say, equalities or inequalities hold for the equivalence classes of indistinguishable stochastic processes; with regard to stochastic processes themselves, these relations are valid only up to an evanescent set.

1.2. Stopping times

Throughout this section, we will assume that a right-continuous stochastic basis $\mathbb{B} = (\Omega, \mathscr{F}, \mathbb{F} = (\mathscr{F}_t)_{t \in \mathbb{R}_+}, \mathsf{P})$ is given.

DEFINITION 1.5.– A *stopping time* is a mapping $T \colon \Omega \to [0, \infty]$ such that $\{T \leqslant t\} \in \mathscr{F}_t$ for every $t \in \mathbb{R}_+$.

It follows from the definition that a stopping time T is an extended (i.e. with values in the extended real line) random variable.

EXAMPLE 1.3.– Let $t \in [0, \infty]$. Put $T(\omega) \equiv t$. Then T is a stopping time.

EXERCISE 1.6.– Let T be a stopping time. Prove that $\{T = \infty\} \in \mathscr{F}_\infty$.

EXERCISE 1.7.– Let T be a stopping time and $t \in \mathbb{R}_+$. Show that $T + t$ is a stopping time. Construct an example of a stopping time $T \geqslant 1$ such that $T - \varepsilon$ is not a stopping time for all $\varepsilon > 0$.

In the next two propositions, the right-continuity of the filtration is used.

PROPOSITION 1.2.– A mapping $T \colon \Omega \to [0, \infty]$ is a stopping time if and only if $\{T < t\} \in \mathscr{F}_t$ for every $t \in \mathbb{R}_+$.

PROPOSITION 1.3.– Let $T_1, \ldots, T_n, \ldots$ be stopping times. Then $\sup_n T_n$ and $\inf_n T_n$ are stopping times.

EXERCISE 1.8.– Prove propositions 1.2 and 1.3.

DEFINITION 1.6.– Let T be a stopping time. Denote by $\mathscr{F}_T$ the following class of sets:

$$\mathscr{F}_T := \{A \in \mathscr{F}_\infty \colon A \cap \{T \leqslant t\} \in \mathscr{F}_t \text{ for every } t \in \mathbb{R}_+\}.$$

EXERCISE 1.9.– Prove that $\mathscr{F}_T$ is a σ-algebra.

The σ-algebra $\mathscr{F}_T$ is usually interpreted as the class of events occurring before or at a random time T. $\mathscr{F}_T$ should not be confused with the smallest σ-algebra, with respect to which the mapping T is measurable. The latter, as a rule, is much more narrow.

Sometimes in the literature, in the definition of $\mathscr{F}_T$, the assumption $A \in \mathscr{F}_\infty$ is replaced by $A \in \mathscr{F}$. The difference between these definitions is not essential.

PROPOSITION 1.4.– Let T be a stopping time. If $A \in \mathscr{F}_T$, then

$$A \cap \{T < t\} \in \mathscr{F}_t \quad \text{for every } t \in \mathbb{R}_+. \tag{1.1}$$

Conversely, if $A \in \mathscr{F}_\infty$ and [1.1] is valid, then $A \in \mathscr{F}_T$.

PROOF.– Let $A \in \mathscr{F}_T$, $t \in \mathbb{R}_+$. Then

$$A \cap \{T < t\} = \bigcup_{k=1}^{\infty} \left(A \cap \{T \leqslant t - 1/k\}\right) \in \mathscr{F}_t.$$

Conversely, let $A \in \mathscr{F}_\infty$ and [1.1] be valid. Then, for any $t \in \mathbb{R}_+$, and for any natural m,

$$A \cap \{T \leqslant t\} = \bigcap_{k=m}^{\infty} \left(A \cap \{T < t + 1/k\}\right) \in \mathscr{F}_{t+1/m},$$

hence,

$$A \cap \{T \leqslant t\} \in \bigcap_{m=1}^{\infty} \mathscr{F}_{t+1/m} = \mathscr{F}_t$$

in view of the right-continuity of the filtration.   □

THEOREM 1.1.– Let T and S be stopping times. Then:

1) if $T \equiv t, t \in [0, \infty]$, then $\mathscr{F}_T = \mathscr{F}_t$;

2) the extended random variable T is $\mathscr{F}_T$-measurable;

3) if $B \in \mathscr{F}_S$, then $B \cap \{S \leqslant T\} \in \mathscr{F}_T$;

4) if $S \leqslant T$, then $\mathscr{F}_S \subseteq \mathscr{F}_T$;

5) $\mathscr{F}_{S \wedge T} = \mathscr{F}_S \cap \mathscr{F}_T$, $\mathscr{F}_{S \vee T} = \mathscr{F}_S \bigvee \mathscr{F}_T$;

6) the sets $\{S < T\}$, $\{S = T\}$ and $\{S > T\}$ belong to $\mathscr{F}_S \cap \mathscr{F}_T$.

PROOF.– The proof of (1) and (2) is left to the reader as an exercise. Let us prove (3). We have, for every $t \in \mathbb{R}_+$,

$$B \cap \{S \leqslant T\} \cap \{T \leqslant t\} = (B \cap \{S \leqslant t\}) \cap \{T \leqslant t\} \cap \{S \wedge t \leqslant T \wedge t\}.$$

Now note that $B \cap \{S \leqslant t\} \in \mathscr{F}_t$ by the definition of $\mathscr{F}_S$, $\{T \leqslant t\} \in \mathscr{F}_t$ by the definition of a stopping time, and $\{S \wedge t \leqslant T \wedge t\} \in \mathscr{F}_t$, because $S \wedge t$ and $T \wedge t$ are $\mathscr{F}_t$-measurable (which follows from the definition of the stopping time).

Assertion (4) follows directly from (3).

The inclusion $\mathscr{F}_{S \wedge T} \subseteq \mathscr{F}_S \cap \mathscr{F}_T$ follows from (4). Let $B \in \mathscr{F}_S \cap \mathscr{F}_T$. Then, for $t \in \mathbb{R}_+$,

$$B \cap \{S \wedge T \leqslant t\} = B \cap (\{S \leqslant t\} \cup \{T \leqslant t\})$$
$$= (B \cap \{S \leqslant t\}) \cup (B \cap \{T \leqslant t\}) \in \mathscr{F}_t.$$

The inclusion $\mathscr{F}_S \bigvee \mathscr{F}_T \subseteq \mathscr{F}_{S \vee T}$ follows from (4). Let $B \in \mathscr{F}_{S \vee T}$. Then $B \cap \{S \leqslant T\} = B \cap \{S \vee T \leqslant T\} \in \mathscr{F}_T$ due to (3). Similarly, $B \cap \{T \leqslant S\} \in \mathscr{F}_S$. Therefore, $B = (B \cap \{S \leqslant T\}) \cup (B \cap \{T \leqslant S\}) \in \mathscr{F}_S \bigvee \mathscr{F}_T$.

Since $\{S > T\} = \Omega \backslash \{S \leqslant T\} \in \mathscr{F}_T$ due to (3) and $\{S = T\} = \{S \geqslant T\} \backslash \{S > T\}$, to prove (6), it is enough to check that $\{S \geqslant T\} \in \mathscr{F}_T$. But this follows from the equality $\{S \geqslant T\} = \{S \wedge T = T\}$ and $\mathscr{F}_T$-measurability of random variables $S \wedge T$ and T, see (2) and (4).   □

EXERCISE 1.10.– Prove assertions (1) and (2) of theorem 1.1.

Given a mapping $T: \Omega \to [0, \infty]$ and a set $A \subseteq \Omega$, define a new mapping $T_A: \Omega \to [0, \infty]$ by

$$T_A(\omega) = \begin{cases} T(\omega), & \text{if } \omega \in A, \\ \infty, & \text{if } \omega \notin A. \end{cases}$$

PROPOSITION 1.5.– Let T be a stopping time, $A \subseteq \Omega$. Then T_A is a stopping time if and only if $A \cap \{T < \infty\} \in \mathscr{F}_T$. In particular, if $A \in \mathscr{F}_T$, then T_A is a stopping time.

PROOF.– The claim follows from

$$\{T_A \leqslant t\} = A \cap \{T \leqslant t\}, \quad t \in \mathbb{R}_+. \qquad \square$$

The stopping time T_A is called the *restriction* of a stopping time T on the set A.

We associate another σ-algebra $\mathscr{F}_{T-}$ with a stopping time T. Elements of $\mathscr{F}_{T-}$ are usually interpreted as events occurring strictly before a stopping time T.

DEFINITION 1.7.– Let T be a stopping time. Denote by $\mathscr{F}_{T-}$ the smallest σ-algebra containing the σ-algebra $\mathscr{F}_0$ and all sets of the form

$$A \cap \{t < T\}, \quad t \in \mathbb{R}_+, \quad A \in \mathscr{F}_t.$$

THEOREM 1.2.– Let T and S be stopping times. Then:

1) if $T \equiv t, t \in [0, \infty]$, then $\mathscr{F}_{T-} = \mathscr{F}_{t-}$;

2) $\mathscr{F}_{T-} \subseteq \mathscr{F}_T$;

3) the extended random variable T is $\mathscr{F}_{T-}$-measurable;

4) if $S \leqslant T$, then $\mathscr{F}_{S-} \subseteq \mathscr{F}_{T-}$;

5) if $B \in \mathscr{F}_S$, then $B \cap \{S < T\} \in \mathscr{F}_{T-}$.

PROOF.– Assertion (1) follows directly from the definitions. Let us prove (2). Since $\mathscr{F}_0 \subseteq \mathscr{F}_T$, it is enough to check that $A \cap \{t < T\} \in \mathscr{F}_T$ if $t \in \mathbb{R}_+$ and $A \in \mathscr{F}_t$. Take $s \in \mathbb{R}_+$ and show that $A \cap \{t < T \leqslant s\} \in \mathscr{F}_s$. Indeed, this set is empty, if $s \leqslant t$; and if $s > t$, then $A \in \mathscr{F}_s$ and $\{t < T \leqslant s\} \in \mathscr{F}_s$.

To prove (3), it is enough to note that $\{t < T\} \in \mathscr{F}_{T-}$ for every $t \in \mathbb{R}_+$.

Let $t \in \mathbb{R}_+, A \in \mathscr{F}_t, S \leqslant T$. Then

$$B := A \cap \{t < S\} \in \mathscr{F}_t.$$

Hence,

$$B = B \cap \{t < T\} \in \mathscr{F}_{T-},$$

and (4) follows.

Now we prove (5). Let $B \in \mathscr{F}_S$. We have

$$B \cap \{S < T\} = \bigcup_{r \in \mathbb{Q}_+} (B \cap \{S \leqslant r\} \cap \{r < T\}),$$

where the union is taken over a *countable* set of rational numbers r. By the definition of $\mathscr{F}_S$, $B \cap \{S \leqslant r\} \in \mathscr{F}_r$. Hence, $B \cap \{S \leqslant r\} \cap \{r < T\} \in \mathscr{F}_{T-}$. The claim follows.   □

PROPOSITION 1.6.– If T is a stopping time and $A \in \mathscr{F}_\infty$, then $A \cap \{T = \infty\} \in \mathscr{F}_{T-}$.

PROOF.– It is easy to see that sets A with the indicated properties constitute a σ-algebra. Therefore, it is enough to check the assertion if $A \in \mathscr{F}_t$, $t \in \mathbb{R}_+$. But in this case

$$A \cap \{T = \infty\} = \bigcap_{n=1}^{\infty} (A \cap \{t + n < T\}) \in \mathscr{F}_{T-}.$$ □

LEMMA 1.1.– Let S and T be stopping times, $S \leqslant T$ and $S < T$ on the set $\{0 < T < \infty\}$. Then $\mathscr{F}_S \subseteq \mathscr{F}_{T-}$.

PROOF.– Let $A \in \mathscr{F}_S$. It follows from the hypotheses that

$$A = (A \cap \{T = 0\}) \cup (A \cap \{S < T\}) \cup (A \cap \{T = \infty\}).$$

Since $\mathscr{F}_S \subseteq \mathscr{F}_T$ by theorem 1.1 (4), we have $A \cap \{T = 0\} \in \mathscr{F}_0 \subseteq \mathscr{F}_{T-}$. Next, $A \cap \{S < T\} \in \mathscr{F}_{T-}$ by theorem 1.2 (5). Finally, $A \cap \{T = \infty\} \in \mathscr{F}_{T-}$ by proposition 1.6.   □

THEOREM 1.3.– Let (T_n) be a monotone sequence of stopping times and $T = \lim_n T_n$:

1) if (T_n) is an increasing sequence, then

$$\mathscr{F}_{T-} = \bigvee_{n=1}^{\infty} \mathscr{F}_{T_n-};$$

moreover, if $\{0 < T < \infty\} \subseteq \{T_n < T\}$ for every n, then

$$\mathscr{F}_{T-} = \bigvee_{n=1}^{\infty} \mathscr{F}_{T_n}.$$

2) if (T_n) is a decreasing sequence, then

$$\mathscr{F}_T = \bigcap_{n=1}^{\infty} \mathscr{F}_{T_n};$$

moreover, if $\{0 < T < \infty\} \subseteq \{T < T_n\}$ for every n, then

$$\mathscr{F}_T = \bigcap_{n=1}^{\infty} \mathscr{F}_{T_n-}.$$

PROOF.–

1) by theorem 1.2 (4), $\mathscr{F}_{T-} \supseteq \bigvee_{n=1}^{\infty} \mathscr{F}_{T_n-}$. To prove the converse inclusion, it is enough to show that all elements that generate the σ-algebra $\mathscr{F}_{T-}$ belong to $\bigvee_{n=1}^{\infty} \mathscr{F}_{T_n-}$. This is obvious for sets from $\mathscr{F}_0$. Let $A \in \mathscr{F}_t, t \in \mathbb{R}_+$. Then

$$A \cap \{t < T\} = \bigcup_{n=1}^{\infty} (A \cap \{t < T_n\}) \in \bigvee_{n=1}^{\infty} \mathscr{F}_{T_n-},$$

because $A \cap \{t < T_n\} \in \mathscr{F}_{T_n-}$.

The second part of the assertion follows from the first one and lemma 1.1;

2) by theorem 1.1 (4), $\mathscr{F}_T \subseteq \bigcap_{n=1}^{\infty} \mathscr{F}_{T_n}$. Let $A \in \bigcap_{n=1}^{\infty} \mathscr{F}_{T_n}$. Fix $t \in \mathbb{R}_+$. Then, for every n, $A \cap \{T_n < t\} \in \mathscr{F}_t$ by proposition 1.4, hence,

$$A \cap \{T < t\} = \bigcup_{n=1}^{\infty} (A \cap \{T_n < t\}) \in \mathscr{F}_t.$$

Therefore, $A \in \mathscr{F}_T$ by proposition 1.4.

It follows from the assumption $\{0 < T < \infty\} \subseteq \{T < T_n\}$ that $\{0 < T_n < \infty\} \subseteq \{T < T_n\}$. Thus, the second part of the assertion follows from the first one and lemma 1.1. □

REMARK 1.2.– Nowhere in this section was the completeness of the stochastic basis used. Using the completeness, we can slightly weaken the assumptions in some statements. Thus, in theorems 1.1 (4) and 1.2 (4), we can assume that $S \leqslant T$ a.s. In lemma 1.1, it is enough to assume that $S \leqslant T$ a.s. and $S < T$ on the set $\{0 < T < \infty\}$ a.s. (the latter means that $\mathsf{P}(S \geqslant T, \ 0 < T < \infty) = 0$). We can also modify assumptions of theorem 1.3 in a corresponding way. All this can be proved either directly or using the statement in the next exercise.

EXERCISE 1.11.– Let T be a stopping time and S be a mapping from Ω to $[0, \infty]$ such that $\{S \neq T\} \in \mathscr{F}$ and $\mathsf{P}(S \neq T) = 0$. Prove that S is a stopping time, $\mathscr{F}_S = \mathscr{F}_T$ and $\mathscr{F}_{S-} = \mathscr{F}_{T-}$.

1.3. Measurable, progressively measurable, optional and predictable σ-algebras

In this section, we introduce four σ-algebras on the product space $\Omega \times \mathbb{R}_+$. We will assume that a right-continuous stochastic basis $\mathbb{B} = (\Omega, \mathscr{F}, \mathbb{F} = (\mathscr{F}_t)_{t \in \mathbb{R}_+}, \mathsf{P})$ is given.

DEFINITION 1.8.– The σ-algebra of *measurable* sets on $\Omega \times \mathbb{R}_+$ is the product $\mathscr{F} \otimes \mathscr{B}(\mathbb{R}_+)$ of the σ-algebra $\mathscr{F}$ and the Borel σ-algebra on $\mathbb{R}_+$. A stochastic process $X = (X_t)_{t \in \mathbb{R}_+}$ with values in E is *measurable* if the mapping $(\omega, t) \rightsquigarrow X(\omega, t)$ is measurable as a mapping from $(\Omega \times \mathbb{R}_+, \mathscr{F} \otimes \mathscr{B}(\mathbb{R}_+))$ to $(E, \mathscr{E})$.

Note that it is not necessary to assume that X is a stochastic process. Indeed, if a mapping $X \colon \Omega \times \mathbb{R}_+ \to E$ is measurable, then the mapping $\omega \rightsquigarrow X_t(\omega) = X(\omega, t)$ is $\mathscr{F}$-measurable for every $t \in \mathbb{R}_+$, i.e. $X = (X_t)$ is a stochastic process in accordance with the definition in section 1.1. Here we use a well-known fact that if a function of two variables is measurable with respect to the product of σ-algebras, then it is measurable in each variable, another one being fixed.

PROPOSITION 1.7.– Let X be a measurable stochastic process with values in E, H be a nonnegative random variable. Then the mapping $X_H \colon \Omega \to E$ defined by $X_H(\omega) = X(\omega, H(\omega))$, is a measurable mapping from $(\Omega, \mathscr{F})$ to $(E, \mathscr{E})$.

PROOF.– X_H is a composition of measurable mappings

$$\omega \rightsquigarrow (\omega, H(\omega)) \quad \text{from} \quad (\Omega, \mathscr{F}) \quad \text{to} \quad (\Omega \times \mathbb{R}_+, \mathscr{F} \otimes \mathscr{B}(\mathbb{R}_+))$$

and

$$(\omega, t) \rightsquigarrow X(\omega, t) \quad \text{from} \quad (\Omega \times \mathbb{R}_+, \mathscr{F} \otimes \mathscr{B}(\mathbb{R}_+)) \quad \text{to} \quad (E, \mathscr{E}). \qquad \square$$

In what follows, the notation X_H will be permanently used but in a wider context, where H may also take the value $+\infty$. In such a case, there appears to be a problem in defining $X(\omega, H(\omega))$ if $H(\omega) = \infty$. We will use the following two solutions (for simplicity, $E = \mathbb{R}$):

1) We denote by $X_H \mathbb{1}_{\{H<\infty\}}$, the random variable which takes value $X(\omega, H(\omega))$, if $H(\omega) < \infty$, and value 0 if $H(\omega) = \infty$. If a process X is measurable and H is an extended random variable with values in $[0, \infty]$, then $X_H \mathbb{1}_{\{H<\infty\}}$ is a random variable.

2) Let us assume additionally that $X = Y$ or $X = Y_-$, where Y is a càdlàg process, there exists a (maybe, infinite) limit $\lim_{t\to\infty} Y_t(\omega)$ for almost all ω. In this case, in section 1.1, we defined random variables Y_∞ and $Y_{\infty-}$, equal to each other, and it is natural to assign the same value to X_∞. Then $X_H(\omega) = X(\omega, H(\omega))$ is defined (uniquely up to a null set) and is an extended random variable (if H is an extended random variable with values in $[0, \infty]$).

The σ-algebra of measurable sets is not connected with the filtration $\mathbb{F}$. It is natural to distinguish processes, whose measurability in ω or in (ω, t) agrees with time stream.

DEFINITION 1.9.– A stochastic process $X = (X_t)_{t\in\mathbb{R}_+}$ with values in E is called *adapted* (relative to the filtration $\mathbb{F}$) if, for every $t \in \mathbb{R}_+$, the mapping $\omega \rightsquigarrow X_t(\omega)$ is measurable as a mapping from $(\Omega, \mathscr{F}_t)$ to $(E, \mathscr{E})$.

If the stochastic basis is complete, if X is adapted and Y is a modification of X, then Y is also adapted.

DEFINITION 1.10.– A stochastic process $X = (X_t)_{t\in\mathbb{R}_+}$ with values in E is called *progressively measurable* if, for every $t \in \mathbb{R}_+$, the mapping $(\omega, s) \rightsquigarrow X(\omega, s)$ is measurable as a mapping from $(\Omega \times [0,t], \mathscr{F}_t \otimes \mathscr{B}([0,t]))$ to $(E, \mathscr{E})$. A set $B \subseteq \Omega \times \mathbb{R}_+$ is called *progressively measurable* if its indicator is a progressively measurable stochastic process, i.e., if, for every $t \in \mathbb{R}_+$,

$$B \cap (\Omega \times [0,t]) \in \mathscr{F}_t \otimes \mathscr{B}([0,t]).$$

EXERCISE 1.12.– Show that a progressively measurable set is measurable. Show that the family of all progressively measurable sets is a σ-algebra. Show that a stochastic process $X = (X_t)_{t\in\mathbb{R}_+}$ with values in E is progressively measurable if and only if the mapping $(\omega, t) \rightsquigarrow X(\omega, t)$ is measurable as a mapping from $\Omega \times \mathbb{R}_+$ to E, where $\Omega \times \mathbb{R}_+$ is equipped with the σ-algebra of progressively measurable sets.

Thus, we have introduced the σ-algebra of progressively measurable sets in $\Omega \times \mathbb{R}_+$, which is contained in the σ-algebra of measurable sets. Correspondingly, a progressively measurable process is measurable. By the above mentioned fact (see also theorem 1.4 below) a progressively measurable process is adapted. As is indicated in the next exercise, a measurable adapted stochastic process may not be progressively measurable.

EXERCISE 1.13.– Let a probability space $(\Omega, \mathscr{F}, \mathsf{P})$ and a stochastic process Y be the same as in example 1.2. Assume that, for every $t \in \mathbb{R}_+$, the σ-algebra $\mathscr{F}_t$ consists of

(Lebesgue measurable) sets of Lebesgue measure 0 or 1. Show that Y is measurable and adapted, but is not progressively measurable.

REMARK 1.3.– If $\mathscr{F}_t \equiv \mathscr{F}$, the progressively measurable and the measurable σ-algebras coincide.

Using this remark, we can often obtain assertions concerning measurable processes as special cases of assertions concerning progressively measurable processes. For instance, proposition 1.7 is a corollary of theorem 1.4.

THEOREM 1.4.– Let X be a progressively measurable stochastic process and T be a stopping time. Then the random variable $X_T \mathbb{1}_{\{T<\infty\}}$ is $\mathscr{F}_T$-measurable. If a limit $\lim_{t\to\infty} X_t(\omega)$ exists for almost all ω, then the random variable X_T is $\mathscr{F}_T$-measurable.

PROOF.– Under the assumptions of the second assertion, the random variable X_∞ is $\mathscr{F}_\infty$-measurable (see exercise 1.5). So the second assertion follows from the first assertion and proposition 1.6.

Fix $t \in \mathbb{R}_+$ and define two mappings:

$$\omega \rightsquigarrow (\omega, t \wedge T(\omega)) \qquad \text{from} \quad \Omega \quad \text{to} \quad \Omega \times [0,t]$$

and

$$(\omega, s) \rightsquigarrow X_s(\omega) \qquad \text{from} \quad \Omega \times [0,t] \quad \text{to} \quad \mathbb{R}.$$

The first mapping is measurable if we take the σ-algebra $\mathscr{F}_t$ on Ω, and if $\Omega \times [0,t]$ is equipped with the σ-algebra $\mathscr{F}_t \otimes \mathscr{B}([0,t])$. Indeed, let $A \in \mathscr{F}_t$, $B \in \mathscr{B}([0,t])$. Then $\{\omega: \omega \in A, \ t \wedge T(\omega) \in B\} = A \cap \{\omega: t \wedge T(\omega) \in B\} \in \mathscr{F}_t$, because the random variable $t \wedge T$ is $\mathscr{F}_t$-measurable. The second mapping is measurable as a mapping from $(\Omega \times [0,t], \mathscr{F}_t \otimes \mathscr{B}([0,t]))$ to $(\mathbb{R}, \mathscr{B}(\mathbb{R}))$ because X is progressively measurable. Therefore, their composition

$$\omega \rightsquigarrow X_{t \wedge T(\omega)}(\omega)$$

is measurable as a mapping from $(\Omega, \mathscr{F}_t)$ to $(\mathbb{R}, \mathscr{B}(\mathbb{R}))$, i.e. $\{X_{t \wedge T} \in B\} \in \mathscr{F}_t$ for any Borel set B. Thus,

$$\{X_T \mathbb{1}_{\{T<\infty\}} \in B\} \cap \{T \leqslant t\} = \{X_T \in B\} \cap \{T \leqslant t\}$$
$$= \{X_{t \wedge T} \in B\} \cap \{T \leqslant t\} \in \mathscr{F}_t.$$

The definition of the σ-algebra $\mathscr{F}_T$ requires confirmation that $C := \{X_T \mathbb{1}_{\{T<\infty\}} \in B\} \in \mathscr{F}_\infty$. As we have proved, $C \cap \{T \leqslant n\} \in \mathscr{F}_n$, hence,

$C \cap \{T < \infty\} = \bigcup_n (C \cap \{T \leqslant n\}) \in \mathscr{F}_\infty$. It remains to note that the set $C \cap \{T = \infty\}$ either is empty or coincides with $\{T = \infty\} \in \mathscr{F}_\infty$.   □

THEOREM 1.5.– Let X be a right-continuous (or left-continuous) adapted stochastic process. Then X is a progressively measurable process.

PROOF.– Let X be right-continuous. Fix $t \in \mathbb{R}_+$. For a natural n, put

$$X_s^n = \begin{cases} X_{k2^{-n}t}, & \text{if } s \in](k-1)2^{-n}t, k2^{-n}t], \ k = 1, \ldots, 2^n, \\ X_0, & \text{if } s = 0. \end{cases}$$

Then, for a Borel set B,

$$\{(\omega, s) \colon X_s^n(\omega) \in B\} = \left(\{\omega \colon X_0^n(\omega) \in B\} \times \{0\}\right) \bigcup$$

$$\left(\bigcup_{k=1}^{2^n} \left(\{\omega \colon X_{k2^{-n}t}^n(\omega) \in B\} \times](k-1)2^{-n}t, k2^{-n}t]\right)\right) \in \mathscr{F}_t \otimes \mathscr{B}([0,t]).$$

Since

$$\lim_{n \to \infty} X_s^n(\omega) = X_s(\omega) \quad \text{for all } \omega \in \Omega, \ s \in [0,t],$$

the mapping $(\omega, s) \rightsquigarrow X(\omega, s)$ is also measurable as a mapping from $(\Omega \times [0,t], \mathscr{F}_t \otimes \mathscr{B}([0,t]))$ to $(\mathbb{R}, \mathscr{B}(\mathbb{R}))$.

The case of a left-continuous X is considered similarly (see also lemma 1.2).   □

DEFINITION 1.11.– The *optional* σ-algebra on $\Omega \times \mathbb{R}_+$ is the smallest σ-algebra, with respect to which all adapted càdlàg stochastic processes are measurable. The optional σ-algebra is denoted by $\mathscr{O}$. A stochastic process $X = (X_t)_{t \in \mathbb{R}_+}$ with values in E is called *optional* if the mapping $(\omega, t) \rightsquigarrow X(\omega, t)$ is measurable as a mapping from $(\Omega \times \mathbb{R}_+, \mathscr{O})$ to $(E, \mathscr{E})$.

Earlier in the literature, the term "well measurable" was used instead of "optional".

According to theorem 1.5, the optional σ-algebra is contained in the σ-algebra of progressively measurable sets, hence, any optional stochastic process is progressively measurable. The difference between these σ-algebras is not essential (see exercise 1.40 in section 1.6), but it exists. An example of a progressively measurable set which is not optional can be found, for example, in Chapter IV of [DEL 78].

Note also that, if the stochastic basis is complete, then every right-continuous adapted process is optional (see Chapter IV, Theorem 65 [DEL 78]).

LEMMA 1.2.– Let X be a left-continuous adapted stochastic process. Then X is optional.

PROOF.– For a natural n, put

$$X_t^n = \sum_{k=0}^{\infty} X_{k2^{-n}} \mathbb{1}_{\{k2^{-n} \leqslant t < (k+1)2^{-n}\}}.$$

The process X^n is adapted and càdlàg, hence, X^n is optional. Since

$$\lim_{n \to \infty} X_t^n(\omega) = X_t(\omega) \quad \text{for all } \omega, t,$$

the process X is also optional. $\qquad \square$

To change "left" and "right" in the previous proof in order to approximate a right-continuous adapted process by left-continuous adapted processes is not possible: under the corresponding construction (see the proof of theorem 1.5), X^n are not adapted.

DEFINITION 1.12.– The *predictable* σ-algebra on $\Omega \times \mathbb{R}_+$ is the smallest σ-algebra, with respect to which all adapted left-continuous stochastic processes are measurable. The predictable σ-algebra is denoted by $\mathscr{P}$. A stochastic process $X = (X_t)_{t \in \mathbb{R}_+}$ with values in E is called *predictable* if the mapping $(\omega, t) \rightsquigarrow X(\omega, t)$ is measurable as a mapping from $(\Omega \times \mathbb{R}_+, \mathscr{P})$ to $(E, \mathscr{E})$.

According to lemma 1.2, the predictable σ-algebra is included in the optional one. As a rule, they do not coincide.

EXERCISE 1.14.– Construct an example of an optional set which is not predictable.

HINT.– Take a discrete filtration and embed it into a continuous filtration as is explained after definition 1.1.

PROPOSITION 1.8.– Let X be an adapted càdlàg process. Then the process X_- is predictable and ΔX is optional. The process X is predictable if and only if ΔX is predictable.

PROOF.– It is enough to note that X_- is left-continuous and adapted. $\qquad \square$

We successively introduced four σ-algebras on $\Omega \times \mathbb{R}_+$, each subsequent σ-algebra being included in the previous algebras. In what follows, we will deal mostly with the optional and predictable σ-algebras. In particular, our current aim is to provide alternative descriptions of these two σ-algebras.

Stochastic intervals are sets in $\Omega \times \mathbb{R}_+$ of one of the following forms:

$$[\![S,T]\!] := \{(\omega,t) \in \Omega \times \mathbb{R}_+ : S(\omega) \leqslant t \leqslant T(\omega)\},$$

$$[\![S,T[\![:= \{(\omega,t) \in \Omega \times \mathbb{R}_+ : S(\omega) \leqslant t < T(\omega)\},$$

$$]\!]S,T]\!] := \{(\omega,t) \in \Omega \times \mathbb{R}_+ : S(\omega) < t \leqslant T(\omega)\},$$

$$]\!]S,T[\![:= \{(\omega,t) \in \Omega \times \mathbb{R}_+ : S(\omega) < t < T(\omega)\},$$

where S and T are stopping times. A stochastic interval $[\![T,T]\!]$ is denoted by $[\![T]\!]$ and called the *graph* of a stopping time T. Let us emphasize that stochastic intervals are subsets of $\Omega \times \mathbb{R}_+$ and are not subsets of $\Omega \times [0,\infty]$. In particular, our definition of the graph of a stopping time differs from the definition of the graph of a mapping in the theory of functions (if T takes value $+\infty$).

All stochastic intervals are optional sets. Moreover, the following statement takes place.

LEMMA 1.3.– Let S and T be stopping times and ξ be an $\mathscr{F}_S$-measurable random variable. Then the stochastic processes $\xi \mathbb{1}_{[\![S,T]\!]}, \xi \mathbb{1}_{[\![S,T[\![}, \xi \mathbb{1}_{]\!]S,T[\![}$ are optional and the stochastic process $\xi \mathbb{1}_{]\!]S,T]\!]}$ is predictable.

PROOF.– Let $\xi = \mathbb{1}_B$, $B \in \mathscr{F}_S$. Then the process $X := \mathbb{1}_B \mathbb{1}_{[\![S,T[\![}$ is càdlàg and, moreover, adapted. Indeed,

$$\{\omega : X_t(\omega) = 1\} = B \cap \{S \leqslant t < T\} = \left(B \cap \{S \leqslant t\}\right) \cap \{t < T\} \in \mathscr{F}_t$$

for any $t \in \mathbb{R}_+$. Therefore, $X = \xi \mathbb{1}_{[\![S,T[\![}$ is an optional process. In order to prove that $\xi \mathbb{1}_{[\![S,T[\![}$ is an optional process for an arbitrary $\mathscr{F}_S$-measurable random variable ξ, it is enough to represent ξ as the limit of linear combinations of indicators of sets from $\mathscr{F}_S$.

That the process $\xi \mathbb{1}_{]\!]S,T]\!]}$ is predictable is proved similarly.

Two remaining processes are optional due to the relations

$$\xi \mathbb{1}_{[\![S,T]\!]} = \lim_{n\to\infty} \xi \mathbb{1}_{[\![S,T+1/n[\![}, \qquad \xi \mathbb{1}_{]\!]S,T[\![} = \lim_{n\to\infty} \xi \mathbb{1}_{[\![S+1/n,T[\![}. \qquad \square$$

The next corollary follows also from the definition of the predictable σ-algebra.

COROLLARY 1.1.– If T is a stopping time, then the stochastic interval $[\![0,T]\!]$ is predictable.

PROOF.– The set $[\![0, T]\!]$ is the complement of the set $]\!]T, \infty]\!]$, which is predictable by lemma 1.3.   □

DEFINITION 1.13.– Let X be a measurable stochastic process, T be a stopping time. The process X *stopped* at time T is denoted by X^T and defined by

$$X_t^T = X_{t \wedge T}.$$

It follows from proposition 1.7 that this definition is correct.

PROPOSITION 1.9.– Let T be a stopping time and let X be a measurable (respectively progressively measurable, respectively optional, respectively predictable) stochastic process. Then the stopped process X^T is measurable (respectively progressively measurable, respectively optional, respectively predictable).

PROOF.– The measurable case reduces to the case of progressively measurable processes according to remark 1.3. The claim in three other cases follows from the identity

$$X^T = X \mathbb{1}_{[\![0,T]\!]} + X_T \mathbb{1}_{\{T < \infty\}} \mathbb{1}_{]\!]T,\infty]\!]}$$

and the predictability of the processes $\mathbb{1}_{[\![0,T]\!]}$ and $X_T \mathbb{1}_{\{T < \infty\}} \mathbb{1}_{]\!]T,\infty]\!]}$: the first process is predictable by corollary 1.1, and the predictability of the second process follows from lemma 1.3 and theorem 1.4.   □

DEFINITION 1.14.– The *début* of a set $B \subseteq \Omega \times \mathbb{R}_+$ is the mapping $D_B \colon \Omega \to [0, \infty]$ defined by

$$D_B(\omega) = \begin{cases} \inf \{ t \in \mathbb{R}_+ : (\omega, t) \in B \}, & \text{if this set is not empty,} \\ \infty, & \text{otherwise.} \end{cases}$$

THEOREM 1.6.– Assume that the stochastic basis satisfies the usual conditions. Then the début D_B of any progressively measurable set B is a stopping time.

It is essential for this theorem to be true that the stochastic basis satisfies the usual conditions. If the stochastic basis is right-continuous but not complete, then we can assert only that there exists a stopping time T such that the set $\{T \neq D_B\}$ is a subset of a $\mathscr{F}$-null set.

The proof of theorem 1.6 is based on the following (difficult) theorem from the measure theory, which is given without proof and in a simplified form.

THEOREM 1.7.– Let $(\Omega, \mathscr{F}, \mathsf{P})$ be a complete probability space and K be a compact metric space. Denote by π the projection from the space $\Omega \times K$ onto Ω. If B is an element of the product of σ-algebras $\mathscr{F} \otimes \mathscr{B}(K)$, then the projection $\pi(B)$ of the set B on Ω belongs to $\mathscr{F}$.

In this theorem, the completeness of the probability space is also essential.

PROOF OF THEOREM 1.6.– It is obvious that, for any $t \in \mathbb{R}_+$, the set $\{D_B < t\}$ is the projection on Ω of the set $B_t := B \cap [\![0, t[\![$, which is considered as a subset in the product space $\Omega \times [0, t]$. Since B is progressively measurable, B_t belongs to the σ-algebra $\mathscr{F}_t \otimes \mathscr{B}([0, t])$. Since the probability space $(\Omega, \mathscr{F}_t, \mathsf{P}|_{\mathscr{F}_t})$ is complete, $\{D_B < t\} \in \mathscr{F}_t$ by theorem 1.7. By proposition 1.2, D_B is a stopping time.   □

Theorem 1.6 will be repeatedly used in this book, in particular, if $B = \{(\omega, t) : X_t(\omega) \in A\}$, where A is a Borel subset of the real line and X is a progressively measurable process. In this case, the début D_B is also called the hitting time of the set A by the process X; we will use the notation $\inf \{t : X_t \in A\}$ for it. There are special cases of hitting times, where the statement of theorem 1.6 can be proved directly (and the completeness of the stochastic basis is not used in these cases).

PROPOSITION 1.10.–

1) Let X be a right-continuous adapted stochastic process with values in $\mathbb{R}^d$, and let A be an open subset of $\mathbb{R}^d$. Then $T := \inf \{t : X_t \in A\}$ is a stopping time.

2) Let X be a right-continuous adapted stochastic process with values in $[0, \infty]$, whose all trajectories are nondecreasing, $a \in [0, \infty]$. Then $T := \inf \{t : X_t \geqslant a\}$ is a stopping time.

PROOF.–

1) For any $t \in (0, \infty)$,

$$\{T < t\} = \bigcup_{r \in \mathbb{Q},\, 0 \leqslant r < t} \{X_r \in A\} \in \mathscr{F}_t.$$

By proposition 1.2, T is a stopping time.

2) The claim follows from the relation $\{T \leqslant t\} = \{X_t \geqslant a\} \in \mathscr{F}_t, t \in \mathbb{R}_+$.   □

Yet another simple case of theorem 1.6 is left to the reader as an exercise.

EXERCISE 1.15.– Let X be a continuous adapted process with values in $\mathbb{R}^d$ and A be a closed subset of $\mathbb{R}^d$. Show that $T := \inf \{t : X_t \in A\}$ is a stopping time.

THEOREM 1.8.– The σ-algebra $\mathscr{O}$ of optional sets is generated by stochastic intervals of the form $[\![0, T[\![$, where T is an arbitrary stopping time.

PROOF.– Denote by $\mathscr{O}_1$, the σ-algebra generated by stochastic intervals $[\![0, T[\![$, where T is an arbitrary stopping time. By lemma 1.3, $\mathscr{O}_1 \subseteq \mathscr{O}$. To prove the converse inclusion, it is enough to check that an adapted càdlàg process X is $\mathscr{O}_1$-measurable.

First, note that, if S and T are stopping times, then

$$[\![S, T[\![= [\![0, T[\![\setminus [\![0, S[\![\in \mathscr{O}_1.$$

Moreover, if $B \in \mathscr{F}_S$, then the process

$$\mathbb{1}_B \mathbb{1}_{[\![S,T[\![} = \mathbb{1}_{[\![S_B,T[\![}$$

is $\mathscr{O}_1$-measurable because S_B is a stopping time by proposition 1.5. As in the proof of lemma 1.3, we conclude that the process $\xi \mathbb{1}_{[\![S,T[\![}$ is $\mathscr{O}_1$-measurable if ξ is an $\mathscr{F}_S$-measurable random variable.

For natural k, put $S_0^k = 0$ and define recursively

$$S_{n+1}^k = \inf \{t > S_n^k \colon |X_t - X_{S_n^k}| > 2^{-k}\}, \quad n = 0, 1, \ldots.$$

Note that S_{n+1}^k is the hitting time of the open set $\{x \in \mathbb{R} \colon |x| > 2^{-k}\}$ by the process $X - X^{S_n^k}$. By induction, taking into account proposition 1.10 (1) and proposition 1.9, we conclude that S_n^k are stopping times for all n and k. It is clear that the sequence $\{S_n^k\}_{n=0,1,2,\ldots}$ is not decreasing and, moreover, its limit as $n \to \infty$ is equal to ∞ for all ω, which follows easily from the fact that every trajectory $X.(\omega)$ is càdlàg.

Define the process

$$X^k = \sum_{n=0}^{\infty} X_{S_n^k} \mathbb{1}_{[\![S_n^k, S_{n+1}^k[\![}.$$

As we have proved above, X^k is $\mathscr{O}_1$-measurable for every k. It remains to note that $X = \lim_{k \to \infty} X^k$. Indeed, by the definition of S_{n+1}^k, we have $|X - X^k| \leqslant 2^{-k}$ on the stochastic interval $[\![S_n^k, S_{n+1}^k[\![$, and the union of these intervals over n, as we have proved, is $\Omega \times \mathbb{R}_+$. $\qquad\square$

The same construction is used in the proof of the next assertion, which will be repeatedly used in the sequel.

THEOREM 1.9.– Let X be an adapted càdlàg process. There exists a sequence $\{T_n\}$ of stopping times such that

$$\{\Delta X \neq 0\} = \bigcup_n [\![T_n]\!]$$

and

$$[\![T_n]\!] \cap [\![T_m]\!] = \varnothing, \quad m \neq n.$$

Any sequence with these properties is called a sequence of stopping times *exhausting* jumps of the process X.

PROOF.– Define stopping times S_n^k as in the proof of theorem 1.8. Put $B_n^k = \{S_n^k < \infty, \ \Delta X_{S_n^k} \neq 0\}$. By proposition 1.8 and theorem 1.4, $B_n^k \in \mathscr{F}_{S_n^k}$. Hence, by proposition 1.5, $T_n^k := \left(S_n^k\right)_{B_n^k}$ is a stopping time. It is easy to see from the definition of S_n^k that $\{\Delta X \neq 0\} = \bigcup_{k,n}[\![T_n^k]\!]$. Enumerate T_n^k in a sequence $\{T_n'\}$; we have $\{\Delta X \neq 0\} = \bigcup_{n=1}^{\infty}[\![T_n']\!]$. Now put $A_1 = \Omega$ and $A_n = \bigcap_{m=1}^{n-1}\{T_m' \neq T_n'\}$, $n \geqslant 2$. By theorem 1.1 (6), $A_n \in \mathscr{F}_{T_n'}$, hence, by proposition 1.5, $T_n := \left(T_n'\right)_{A_n}$ is a stopping time. Then, $\{T_n\}$ is a required sequence. □

The following theorem provides three characterizations of the predictable σ-algebra. The second characterization is the most useful: the corresponding family of sets is a semiring in $\Omega \times \mathbb{R}_+$.

THEOREM 1.10.– The predictable σ-algebra $\mathscr{P}$ is generated by any one of the following families of sets or processes:

1) $B \times \{0\}$, $B \in \mathscr{F}_0$, and $[\![0,T]\!]$, where T is an arbitrary stopping time;

2) $B \times \{0\}$, $B \in \mathscr{F}_0$, and $B \times]s,t]$, $s,t \in \mathbb{R}_+$, $s < t$, $B \in \mathscr{F}_s$;

3) continuous adapted stochastic processes.

PROOF.– Denote by $\mathscr{P}_i$, $i = 1,2,3$, the σ-algebra generated by the ith family. It is clear that all sets and processes in these families are predictable, hence, $\mathscr{P}_i \subseteq \mathscr{P}$, $i = 1,2,3$.

Sets $B \times]s,t]$, where $s,t \in \mathbb{R}_+$, $s < t$, $B \in \mathscr{F}_s$, can be represented as $[\![0,t_B]\!] \setminus [\![0,s_B]\!]$, therefore, $\mathscr{P}_2 \subseteq \mathscr{P}_1$.

Next, sets $B \times]s,t]$, where $s,t \in \mathbb{R}_+$, $s < t$, $B \in \mathscr{F}_s$, are represented as $\{0 < Y \leqslant t - s\}$ where $Y = (Y_u)_{u \in \mathbb{R}_+}$ is a continuous adapted process given by $Y_u = \mathbb{1}_B(u - s)^+$. Sets $B \times \{0\}$, where $B \in \mathscr{F}_0$, are represented as $\{Y = 1\}$, where

$Y = (Y_u)_{u \in \mathbb{R}_+}$ is a continuous adapted process given by $Y_u = \mathbb{1}_B - u$. Therefore, $\mathscr{P}_2 \subseteq \mathscr{P}_3$.

Finally, let X be a left-continuous adapted process. For a natural n, put

$$X^n = X_0 \mathbb{1}_{\{0\}} + \sum_{k=0}^{\infty} X_{k2^{-n}} \mathbb{1}_{]k2^{-n},(k+1)2^{-n}]}.$$

It is clear that X^n are measurable with respect to $\mathscr{P}_2$, hence, $X = \lim_{n \to \infty} X^n$ is also $\mathscr{P}_2$-measurable. Thus, $\mathscr{P} \subseteq \mathscr{P}_2$. □

The next proposition is complementary to theorem 1.4. It will be used for a characterization of predictable càdlàg processes.

PROPOSITION 1.11.– Let X be a predictable stochastic process and T a stopping time. Then the random variable $X_T \mathbb{1}_{\{T < \infty\}}$ is $\mathscr{F}_{T-}$-measurable.

PROOF.– First, consider two special cases: $X = \mathbb{1}_B \mathbb{1}_{\{0\}}$, $B \in \mathscr{F}_0$, and $X = \mathbb{1}_B \mathbb{1}_{]s,t]}$, $s, t \in \mathbb{R}_+$, $s < t$, $B \in \mathscr{F}_s$.

In the first case, $X_T \mathbb{1}_{\{T < \infty\}} = \mathbb{1}_{B \cap \{T=0\}}$ and $B \cap \{T = 0\} \in \mathscr{F}_0 \subseteq \mathscr{F}_{T-}$.

In the second case, $X_T \mathbb{1}_{\{T < \infty\}} = \mathbb{1}_{B \cap \{s < T \leqslant t\}}$, $B \cap \{s < T\} \in \mathscr{F}_{T-}$ by the definition of the σ-algebra $\mathscr{F}_{T-}$, and $\{T \leqslant t\} \in \mathscr{F}_{T-}$ by theorem 1.2 (3).

Now let $\mathscr{H}$ be the set of predictable stochastic processes X such that the random variable $X_T \mathbb{1}_{\{T < \infty\}}$ is $\mathscr{F}_{T-}$-measurable. It is clear that $\mathscr{H}$ is a linear space, contains constants and is stable under limits in converging sequences. It follows from the monotone class theorem (theorem A.3) that $\mathscr{H}$ is the set of all predictable stochastic processes. Indeed, let $\mathscr{C}$ be the set of processes X corresponding to two cases considered above. Obviously, $\mathscr{C}$ is stable under multiplication and it has been proved that $\mathscr{C} \subseteq \mathscr{H}$. By theorem 1.10, $\mathscr{P} = \sigma\{\mathscr{C}\}$. □

PROPOSITION 1.12.– Assume that the stochastic basis satisfies the usual conditions. Then an evanescent measurable set is predictable.

PROOF.– If a set B is evanescent, then $B \subseteq A \times \mathbb{R}_+$, where $A \in \mathscr{F}$ and $\mathsf{P}(A) = 0$, hence $A \in \mathscr{F}_0$.

Put $\mathscr{C} := \{C \times [t, \infty) \colon C \in \mathscr{F}, \ t \in \mathbb{R}_+\}$. It is clear that $\mathscr{C}$ is a π-system generating the σ-algebra of measurable sets. Now define $\mathscr{D} := \{D \in \mathscr{F} \otimes \mathscr{B}(\mathbb{R}_+) \colon D \cap (A \times \mathbb{R}_+) \in \mathscr{P}\}$. It is easy to see that $\mathscr{D}$ is a λ-system and $\mathscr{C} \subseteq \mathscr{D}$. By theorem A.2 on π-λ-systems, we have $\mathscr{D} = \mathscr{F} \otimes \mathscr{B}(\mathbb{R}_+)$. In particular, $B \in \mathscr{D}$, hence, $B \in \mathscr{P}$ by the definition of $\mathscr{D}$. □

COROLLARY 1.2.– Assume that the stochastic basis satisfies the usual conditions. Let X be a progressively measurable (respectively optional, respectively predictable) stochastic process, and let Y be a measurable process indistinguishable with X. Then Y is progressively measurable (respectively optional, respectively predictable).

EXERCISE 1.16.– Justify the inclusion $\mathscr{C} \subseteq \mathscr{D}$ in the proof of proposition 1.12.

1.4. Predictable stopping times

Let us assume that a right-continuous stochastic basis $\mathbb{B} = (\Omega, \mathscr{F}, \mathbb{F} = (\mathscr{F}_t)_{t \in \mathbb{R}_+}, \mathrm{P})$ is given.

DEFINITION 1.15.– A stopping time T is called *predictable* if the stochastic interval $[\![0, T[\![$ is predictable.

REMARK 1.4.– It is not necessary to assume *a priori* in this definition that T is a stopping time. Indeed, if T is a mapping from Ω to $[0, \infty]$ such that the set $B := \{(\omega, t) \in \Omega \times \mathbb{R}_+ : 0 \leqslant t < T(\omega)\}$ is progressively measurable, then $X := \mathbb{1}_B$ is an adapted process. Hence, $\{X_t = 0\} = \{T \leqslant t\} \in \mathscr{F}_t$ for every $t \in \mathbb{R}_+$, i.e. T is a stopping time.

LEMMA 1.4.– If T is a stopping time and $t > 0$, then $T + t$ is a predictable stopping time.

PROOF.– Indeed,

$$[\![0, T + t[\![= \bigcup_{n > 1/t} [\![0, T + t - 1/n]\!],$$

and the sets on the right are predictable by exercise 1.7 and corollary 1.1.

EXERCISE 1.17.– Construct an example of a stopping time, which is not predictable.

HINT.– Take a discrete filtration and embed it into a continuous one as is explained after definition 1.1.

LEMMA 1.5.– Let T be a stopping time. The following statements are equivalent:

1) T is predictable;

2) $[\![T, \infty[\![\in \mathscr{P}$;

3) $[\![T]\!] \in \mathscr{P}$.

PROOF.– Obviously, (1) and (2) are equivalent. Since the stochastic interval $]T, \infty[$ is predictable by lemma 1.3, it follows from the relations

$$[\![T]\!] = [\![T, \infty[\![\, \setminus \,]\!]T, \infty[\![, \qquad [\![T, \infty[\![\, = \, [\![T]\!] \cup]\!]T, \infty[\![$$

that (2) and (3) are equivalent. □

The next statement complements theorem 1.10 (see also theorem 1.8).

THEOREM 1.11.– The σ-algebra $\mathscr{P}$ of predictable sets is generated by stochastic intervals of the form $[\![0, T[\![$, where T is a predictable stopping time.

PROOF.– Denote by $\mathscr{P}_1$ the σ-algebra generated by intervals $[\![0, T[\![$, where T is a predictable stopping time. By definition, $[\![0, T[\![\, \in \, \mathscr{P}$ for every predictable T, hence, $\mathscr{P}_1 \subseteq \mathscr{P}$.

Let T be an arbitrary stopping time. Then,

$$[\![0, T]\!] = \bigcap_{n=1}^{\infty} [\![0, T + 1/n[\![\, \in \, \mathscr{P}_1,$$

because $T + 1/n$ are predictable by lemma 1.4. Next, let $B \in \mathscr{F}_0$. Put $T(\omega) = 0$, if $\omega \notin B$, and $T(\omega) = \infty$, if $\omega \in B$. Clearly, T is a stopping time, moreover, T is predictable because $[\![0, T[\![\, = \, B \times \mathbb{R}_+ \in \mathscr{P}$. Therefore,

$$B \times \{0\} = [\![0, T[\![\, \cap \, [\![0]\!] \, \in \, \mathscr{P}_1.$$

By theorem 1.10, we get the converse inclusion $\mathscr{P} \subseteq \mathscr{P}_1$. □

PROPOSITION 1.13.– Let $\{T_n\}$ be a sequence of predictable stopping times. Then

1) $T := \sup_n T_n$ is a predictable stopping time;

2) If $S := \inf_n T_n$ and $\bigcup_n \{S = T_n\} = \Omega$, then S is a predictable stopping time.

In particular, the class of predictable stopping times is stable with respect to a finite number of maxima and minima. The infimum of a countable number of predictable stopping times, in general, is not predictable. For example, let S be a stopping time, which is not predictable (see exercise 1.17). Put $T_n = S + 1/n$. Then all T_n are predictable and $S = \inf_n T_n$.

PROOF.– Under the assumptions of the proposition,

$$[\![0, T[\![\, = \, \bigcup_n [\![0, T_n[\![, \qquad [\![0, S[\![\, = \, \bigcap_n [\![0, T_n[\![. \qquad \Box$$

The following statement and the exercise after it complement proposition 1.5.

PROPOSITION 1.14.– Let T be a predictable stopping time and $A \in \mathscr{F}_{T-}$. Then, T_A is a predictable stopping time.

PROOF.– Note that

$$T_{\cup A_n} = \inf_n T_{A_n}, \qquad T_{\cap A_n} = \sup_n T_{A_n}.$$

Hence, by proposition 1.13, the family $\mathscr{G} := \{A \in \mathscr{F}_T : T_A \text{ is predictable}\}$ is closed under countable unions and intersections. Clearly, $\Omega \in \mathscr{G}$. Let $A \in \mathscr{G}$ and $A^c := \Omega \setminus A$. Then

$$[\![T, T_A[\![= [\![T, \infty[\![\setminus [\![T_A, \infty[\![\in \mathscr{P},$$

therefore,

$$[\![0, T_{A^c}[\![= (\Omega \times \mathbb{R}_+) \setminus [\![T, T_A[\![\in \mathscr{P},$$

i.e. $A^c \in \mathscr{G}$.

Thus, $\mathscr{G}$ is a σ-algebra, and it is enough to verify that sets generating the σ-algebra $\mathscr{F}_{T-}$, are in $\mathscr{G}$. Let $A \in \mathscr{F}_0$, then the indicator of $A^c \times [0, \infty[$ is an adapted continuous process, so

$$[\![0, T_A[\![= [\![0, T[\![\cup (A^c \times [0, \infty[) \in \mathscr{P},$$

hence, $A \in \mathscr{G}$. Now let $A = B \cap \{t < T\}$, where $B \in \mathscr{F}_t$. Then $A \in \mathscr{F}_t$ and

$$[\![0, T_{A^c}[\![= [\![0, T[\![\cup (A \times]t, \infty[) \in \mathscr{P}.$$

Consequently, $A^c \in \mathscr{G}$, hence $A \in \mathscr{G}$. □

EXERCISE 1.18.– Let T be a stopping time, $A \in \mathscr{F}_T$, T_A be a predictable stopping time. Show that $A \in \mathscr{F}_{T-}$.

HINT.– Use proposition 1.15 below.

The next statement extends theorem 1.1 (3) and theorem 1.2 (5).

PROPOSITION 1.15.– Let S be a predictable stopping time, $B \in \mathscr{F}_{S-}$, and let T be a stopping time. Then $B \cap \{S \leqslant T\} \in \mathscr{F}_{T-}$.

PROOF.– We have

$$B \cap \{S \leqslant T\} = \{S_B \leqslant T < \infty\} \cup \left(B \cap \{T = \infty\}\right).$$

The second set on the right is in $\mathscr{F}_{T-}$ by proposition 1.6. With regard to the first set, the process $X := \mathbb{1}_{[\![S_B, \infty[\![}$ is predictable due to the previous proposition, hence,

$$\{S_B \leqslant T < \infty\} = \{X_T \mathbb{1}_{\{T < \infty\}} = 1\} \in \mathscr{F}_{T-}$$

by proposition 1.11. □

The next statement follows from theorem 1.2 (5) and proposition 1.15.

COROLLARY 1.3.– Let S be a predictable stopping time, $B \in \mathscr{F}_{S-}$, T a stopping time. Then $B \cap \{S = T\} \in \mathscr{F}_{T-}$.

The next assertion complements lemma 1.3.

LEMMA 1.6.– Let S be a predictable stopping time, Y an $\mathscr{F}_{S-}$-measurable random variable, T a stopping time. Then, the process $Y \mathbb{1}_{[\![S,T]\!]}$ is predictable.

PROOF.– As above, it is enough to consider the case $Y = \mathbb{1}_B$, where $B \in \mathscr{F}_{S-}$. Then

$$Y \mathbb{1}_{[\![S,T]\!]} = \mathbb{1}_{[\![S_B,T]\!]},$$

S_B is predictable by proposition 1.14, and

$$[\![S_B, T]\!] = [\![0, T]\!] \setminus [\![0, S_B[\![\in \mathscr{P}.$$ □

Any stopping time T is the début of a predictable set $]\!]T, \infty[\![$. So additional assumptions are needed for the début of a predictable set to be a predictable stopping time.

PROPOSITION 1.16.– Assume that the stochastic basis $\mathbb{B} = (\Omega, \mathscr{F}, \mathbb{F} = (\mathscr{F}_t)_{t \in \mathbb{R}_+}, \mathsf{P})$ is complete. Let A be a predictable set and let $T := D_A$ be its début. If $[\![T]\!] \subseteq A$, then T is a predictable stopping time.

PROOF.– By theorem 1.6, T is a stopping time. Therefore,

$$[\![T]\!] = A \cap [\![0, T]\!] \in \mathscr{P},$$

and the claim follows from lemma 1.5. □

Two special cases of proposition 1.16 can be proved without using theorem 1.6. They are given as exercises.

EXERCISE 1.19 (see proposition 1.10 (2)).– Let X be a right-continuous predictable stochastic process with values in $[0, \infty]$, whose all trajectories are nondecreasing, $a \in [0, \infty]$. Prove that $T := \inf \{t \colon X_t \geqslant a\}$ is a predictable stopping time.

We should warn the reader against a possible mistake in a situation which is quite typical in stochastic calculus (though it does not appear in this book). For example, let $a = (a_t)$ be a nonnegative progressively measurable (or predictable) stochastic process and $X_t := \int_0^t a_s \, ds$, where the integral is taken pathwise. Put $T := \inf \{t \colon X_t = +\infty\}$. Then the stopping time T may not be predictable. The point is that, though X is left-continuous and, hence, predictable, its continuity and right-continuity may fail at T.

EXERCISE 1.20 (continuation of exercise 1.15).– Let X be a continuous adapted stochastic process with values in $\mathbb{R}^d$ and let A be a closed subset of $\mathbb{R}^d$. Prove that $T := \inf \{t \colon X_t \in A\}$ is a predictable stopping time.

From now on we will assume that the stochastic basis $\mathbb{B} = (\Omega, \mathscr{F}, \mathbb{F} = (\mathscr{F}_t)_{t \in \mathbb{R}_+}, \mathsf{P})$ satisfies the usual conditions.

Let us formulate without proof a difficult theorem on sections. We will use it, in particular, for an alternative description of predictable stopping times. It is supposed that there is given a family $\mathscr{A}$ of stopping times, which satisfies the following assumptions:

A1) $\mathscr{A}$ contains stopping times that are equal to 0 and $+\infty$ identically;

A2) if $T \in \mathscr{A}$, $S = T$ a.s., then $S \in \mathscr{A}$;

A3) if $S, T \in \mathscr{A}$, then $S \wedge T \in \mathscr{A}$ and $S \vee T \in \mathscr{A}$;

A4) if $S, T \in \mathscr{A}$, then $S_{\{S < T\}} \in \mathscr{A}$;

A5) if $\{T_n\}$ is a nondecreasing sequence of stopping times from $\mathscr{A}$, then $\sup_n T_n \in \mathscr{A}$.

THEOREM 1.12.– Let a family $\mathscr{A}$ of stopping times satisfy assumptions (A1)–(A5), $\mathscr{T}$ be a σ-algebra of subsets of the space $\Omega \times \mathbb{R}_+$, which is generated by all stochastic intervals $[\![0, T[\![$, where $T \in \mathscr{A}$. Let π be the projection from the space $\Omega \times \mathbb{R}_+$ onto Ω. For any set $A \in \mathscr{T}$ and for any $\varepsilon > 0$, there is a stopping time $T \in \mathscr{A}$ such that

$$[\![T]\!] \subseteq A$$

and

$$\mathsf{P}(\pi(A)) \leqslant \mathsf{P}(T < \infty) + \varepsilon.$$

Obviously, the set $\pi(A)$ coincides with the set $\{D_A < \infty\}$, where D_A is the début of A.

If $\mathscr{A}$ is the set of *all* stopping times, then, clearly, it satisfies assumptions (A1)–(A5) and, by theorem 1.8, $\mathscr{T} = \mathcal{O}$. So, the following theorem holds.

COROLLARY 1.4 (optional section theorem).– For any optional set A and any $\varepsilon > 0$, there exists a stopping time T such that

$$[\![T]\!] \subseteq A$$

and

$$\mathsf{P}(\pi(A)) \leqslant \mathsf{P}(T < \infty) + \varepsilon.$$

An example of an optional set which does not admit a full section will be given in exercise 1.34, i.e. corollary 1.4 is not valid with $\varepsilon = 0$.

DEFINITION 1.16.– A sequence of stopping times $\{T_n\}$ *foretells* a stopping time T if $\{T_n\}$ is nondecreasing, $T = \lim_n T_n$ and $\{T > 0\} \subseteq \{T_n < T\}$ for every n. A stopping time T is called *foretellable* if there exists a sequence of stopping times which foretells it.

EXERCISE 1.21.– Let T be a foretellable stopping time, and $S = T$ a.s. Show that S is a foretellable stopping time.

LEMMA 1.7.– A foretellable stopping time is predictable.

PROOF.– Let a sequence of stopping times $\{T_n\}$ foretell a stopping time T. Then, by lemma 1.3

$$]\!]0, T[\![= \bigcup_n]\!]0, T_n]\!] \in \mathscr{P}.$$

By proposition 1.14, $0_{\{T>0\}}$ is a predictable stopping time and $[\![0_{\{T>0\}}]\!] \in \mathscr{P}$ by lemma 1.5. Therefore,

$$[\![0, T[\![= [\![0_{\{T>0\}}]\!] \cup]\!]0, T[\![\in \mathscr{P}. \qquad \square$$

EXERCISE 1.22.– Show that the σ-algebra $\mathscr{P}$ of predictable sets is generated by stochastic intervals of the form $[\![0, T[\![$, where T is a foretellable stopping time.

HINT.– Use the same arguments as in the proof of theorem 1.11.

EXERCISE 1.23.– Prove that proposition 1.13 is still valid if predictable stopping times are replaced by foretellable stopping times.

HINTS FOR PROPOSITION 1.13 PART (2).– It is enough to consider the case where $\{T_n\}$ is a decreasing sequence of foretellable stopping times and, for all ω, $T_n(\omega) = S(\omega)$ for n large enough. Let ρ be a metric on $[0, \infty]$, compatible with the natural topology. For every n, we can find a sequence $\{T_{n,p}\}_{p=1,2,\ldots}$ which foretells T_n and such that

$$P(\{\omega\colon \varrho(T_{n,p}(\omega), T_n(\omega)) > 2^{-p}\}) \leqslant 2^{-(n+p)}.$$

Put

$$S_p = \inf_n T_{n,p}.$$

Prove that the sequence $\{S_p\}$ is nondecreasing and $S_p < S$ on the set $\{S > 0\}$ for every p. Prove that, for every p,

$$P(\{\omega\colon \varrho(\lim_p S_p(\omega), S(\omega)) > 2^{-p}\}) \leqslant P(\{\omega\colon \varrho(T_{n,p}(\omega), S(\omega)) > 2^{-p}\})$$

$$\leqslant P(\{\omega\colon \varrho(T_{n,p}(\omega), T_n(\omega)) > 2^{-p}\}) \leqslant 2^{-p}.$$

Conclude that $S = \lim_p S_p$ a.s.

EXERCISE 1.24.– Prove that proposition 1.14 is still valid if predictable stopping times are replaced by foretellable stopping times.

HINTS.– In contrast to the proof of proposition 1.14, define

$$\mathscr{G} := \{A \in \mathscr{F}_T\colon T_A \text{ and } T_{A^c} \text{ are foretellable}\}.$$

Prove that $\mathscr{G}$ is a σ-algebra, for which use exercise 1.23. Next, show that if a sequence of stopping times $\{T_n\}$ foretells a stopping time T, then $\mathscr{F}_{T_n} \subseteq \mathscr{G}$, and use the second assertion of theorem 1.3 (1).

EXERCISE 1.25.– Prove that the family $\mathscr{A}$ of all foretellable stopping times satisfies assumptions (A1)–(A5).

HINT.– Use proposition 1.15 and the previous exercise to verify (A4) .

Now we are in a position to prove two important theorems.

THEOREM 1.13.– A stopping time is predictable if and only if it is foretellable.

THEOREM 1.14 (predictable section theorem).– For any predictable set A and for any $\varepsilon > 0$, there exists a predictable stopping time T such that

$$[\![T]\!] \subseteq A$$

and

$$\mathsf{P}(\pi(A)) \leqslant \mathsf{P}(T < \infty) + \varepsilon.$$

PROOF OF THEOREMS 1.13 AND 1.14.– Theorem 1.14 with a foretellable (and not only with a predictable) stopping time T follows from theorem 1.12 applied to the family $\mathscr{A}$ of all foretellable stopping times. This is possible due to exercises 1.22 and 1.25.

To prove theorem 1.13, due to lemma 1.7, we need to check only the necessity. Thus, let T be a predictable stopping time. If $T = \infty$ a.s., then T, obviously, is foretellable, so we may suppose that $\mathsf{P}(T < \infty) > 0$. By lemma 1.5, the graph $[\![T]\!]$ is a predictable set. Therefore, by the predictable section theorem, there is a sequence $\{T_n\}$ of foretellable stopping times with the graphs containing in the graph of T, and such that

$$\mathsf{P}(T < \infty) \leqslant \mathsf{P}(T_n < \infty) + 1/n.$$

Replacing T_n by $T_1 \wedge \cdots \wedge T_n$, we may assume that the sequence $\{T_n\}$ is decreasing. Therefore, $\lim_n T_n$ is a foretellable stopping time by exercise 1.23. It remains to note that $T = \lim_n T_n$ a.s. □

The following statement is a typical application of the optional and predictable section theorems.

THEOREM 1.15.– Let X and Y be optional (respectively predictable) nonnegative or bounded stochastic processes. If for arbitrary (respectively arbitrary predictable) stopping time T,

$$\mathsf{E}X_T \mathbb{1}_{\{T<\infty\}} = \mathsf{E}Y_T \mathbb{1}_{\{T<\infty\}}, \qquad\qquad [1.2]$$

then X and Y are indistinguishable.

PROOF.– We consider the case of predictable processes, the proof in the optional case is similar. If the processes X and Y are not indistinguishable, then at least one of the predictable sets

$$\{(\omega, t) \colon X_t(\omega) > Y_t(\omega)\} \qquad \text{and} \qquad \{(\omega, t) \colon X_t(\omega) < Y_t(\omega)\}$$

is not evanescent. Then it follows from theorem 1.14 that there exists a predictable stopping time T such that the equality [1.2] does not hold. □

REMARK 1.5.– In order to apply theorem 1.15, it is necessary to verify [1.2] for any stopping time, *finite or not* (see remark 1.7).

1.5. Totally inaccessible stopping times

Unless otherwise stated, we shall assume that a stochastic basis $\mathbb{B} = (\Omega, \mathscr{F}, \mathbb{F} = (\mathscr{F}_t)_{t \in \mathbb{R}_+}, \mathsf{P})$ satisfying the usual conditions is given.

DEFINITION 1.17.– A stopping time T is called *totally inaccessible* if $\mathsf{P}(\omega : T(\omega) = S(\omega) < \infty) = 0$ (i.e. the set $[\![T]\!] \cap [\![S]\!]$ is evanescent) for any predictable stopping time S.

EXERCISE 1.26.– Let a mapping $T \colon \Omega \to [0, \infty]$ be given; moreover, $\{T = \infty\} \in \mathscr{F}$ and $\mathsf{P}(T = \infty) = 1$. Show that T is a predictable and totally inaccessible stopping time. Prove that there are no other stopping times that are predictable and totally inaccessible simultaneously.

HINT.– Use lemma 1.7.

DEFINITION 1.18.– A stopping time T is called *accessible* if there exists a sequence $\{T_n\}$ of predictable stopping times such that

$$[\![T]\!] \subseteq \bigcup_n [\![T_n]\!] \tag{1.3}$$

up to an evanescent set, i.e.,

$$\mathsf{P}\left(\bigcup_n \{\omega : T_n(\omega) = T(\omega) < \infty\}\right) = \mathsf{P}(T < \infty).$$

Under these conditions, we say that the sequence $\{T_n\}$ *embraces* T.

REMARK 1.6.– Due to exercise 1.26, the words "up to an evanescent set" in definition 1.18 can be omitted.

It is clear that a predictable stopping time is accessible. In exercise 1.33 we will construct an example of an accessible stopping time T, which is not predictable; moreover, in this example, if $\{T_n\}$ embraces T, then the inclusion in [1.3] is strict, i.e. the set

$$\left(\bigcup_n [\![T_n]\!]\right) \setminus [\![T]\!]$$

is not evanescent. It is useful to note, however, that, among all sequences $\{T_n\}$ embracing an attainable stopping time T, we can find a sequence such that the set $\cup_n [\![T_n]\!]$ is minimal up to an evanescent set (see exercise 1.41 in the next section).

We know that stochastic intervals $[\![0, T[\![$ generate the optional σ-algebra, if T runs over the class of all stopping times, and generate the predictable σ-algebra, if T runs over the class of all predictable stopping times. The σ-algebra generated by stochastic intervals $[\![0, T[\![$, where T runs over the class of all accessible stopping times, is called the *accessible* σ-algebra. It contains the predictable σ-algebra, is included into the optional one, and, in general, differs from both of them. The importance of the accessible σ-algebra is not as high as that of the optional and predictable σ-algebras.

The next theorem shows that every stopping time can be decomposed into accessible and totally inaccessible parts.

THEOREM 1.16.– Let T be a stopping time. There exists a unique (up to a null set) decomposition of the set $\{T < \infty\}$ into two sets C and B from the σ-algebra $\mathscr{F}_{T-}$ such that T_C is accessible and T_B is totally inaccessible. The stopping times T_C and T_B are called the accessible and totally inaccessible parts, respectively, of the stopping time T.

PROOF.– Denote by $\mathscr{A}$ the family of sets of the form

$$\left(\bigcup_n \{T_n = T < \infty\} \right),$$

where $\{T_n\}$ is an arbitrary sequence of predictable stopping times. By corollary 1.3, $\mathscr{A} \subseteq \mathscr{F}_{T-}$. It is clear that $\mathscr{A}$ is stable under taking countable unions. Hence,

$$\sup_{A \in \mathscr{A}} \mathsf{P}(A)$$

is attained at a set, say, C from $\mathscr{A}$ (in other words, $C = \text{ess sup}\{A \colon A \in \mathscr{A}\}$). Obviously, the stopping time T_C is accessible. Put $B = \{T < \infty\} \setminus C$. Let S be a predictable stopping time. Then

$$C \cup \{T_B = S < \infty\} = C \cup \{T = S < \infty\} \in \mathscr{A},$$

$$C \cap \{T_B = S < \infty\} = \varnothing,$$

therefore, $\mathsf{P}(T_B = S < \infty) = 0$ by the definition of C. Thus, T_B is totally inaccessible. The uniqueness of the decomposition is proved similarly. $\qquad\square$

Theorem 1.16 implies,

COROLLARY 1.5.– A stopping time T is accessible (respectively totally inaccessible) if and only if

$$P(\omega \colon T(\omega) = S(\omega) < \infty) = 0$$

for every totally inaccessible (respectively accessible) stopping time S.

Theorem 1.16 helps us to prove the following theorem, which plays an important role in the rest of the book.

THEOREM 1.17.– Let X be an adapted càdlàg process. There exists a sequence $\{T_n\}$ of stopping times such that every T_n is either predictable or totally inaccessible,

$$\{\Delta X \neq 0\} \subseteq \bigcup_n [\![T_n]\!]$$

and

$$[\![T_n]\!] \cap [\![T_m]\!] = \varnothing, \quad m \neq n.$$

PROOF.– It follows from theorems 1.9 and 1.16, and remark 1.6, that there exists a sequence $\{S_n\}$ of stopping times which meets the requirements of the theorem except that the graphs are disjoint. As in the proof of theorem 1.9, put $A_1 = \Omega$ and $A_n = \bigcap_{m=1}^{n-1}\{S_m \neq S_n\}$, $n \geqslant 2$, $T_n := (S_n)_{A_n}$; every T_n is a stopping time. Moreover, if S_n is totally inaccessible, then T_n is totally inaccessible by the definition. If S_n is predictable, then $A_n \in \mathscr{F}_{S_n-}$ by corollary 1.3, hence, T_n is predictable by proposition 1.14. Therefore, $\{T_n\}$ is a required sequence. □

THEOREM 1.18.– Let X be a predictable càdlàg process. There exists a sequence $\{T_n\}$ of predictable stopping times such that

$$\{\Delta X \neq 0\} = \bigcup_n [\![T_n]\!]$$

and

$$[\![T_n]\!] \cap [\![T_m]\!] = \varnothing, \quad m \neq n.$$

PROOF.– By the previous theorem, there exists a sequence $\{S_n\}$ of stopping times such that every S_n is either predictable or totally inaccessible,

$$\{\Delta X \neq 0\} \subseteq \bigcup_n [\![S_n]\!]$$

and

$$[\![S_n]\!] \cap [\![S_m]\!] = \varnothing, \quad m \neq n.$$

Put

$$B = \{\Delta X \neq 0\} \setminus \left(\bigcup_{n \,:\, S_n \text{ predictable}} [\![S_n]\!] \right).$$

By proposition 1.8, the process ΔX is predictable, hence, $B \in \mathscr{P}$. Let S be a predictable stopping time such that $[\![S]\!] \subseteq B$. Then

$$[\![S]\!] \subseteq \left(\bigcup_{n \,:\, S_n \text{ totally inaccessible}} [\![S_n]\!] \right).$$

Hence, $P(S = \infty) = 1$. By the predictable section theorem, B is an evanescent set. Now put

$$T_n = (S_n)_{\{S_n < \infty, \Delta X_{S_n} \neq 0\}}.$$

If S_n is predictable, then T_n is predictable by propositions 1.11 and 1.14 (or because $[\![T_n]\!] = [\![S_n]\!] \cap \{\Delta X \neq 0\} \in \mathscr{P}$). If S_n is totally inaccessible, then $[\![T_n]\!] \subseteq B$. As it has been proved, $P(T_n < \infty) = 0$, Thus, T_n is predictable by exercise 1.26. So, $\{T_n\}$ is a required sequence. □

THEOREM 1.19.– A càdlàg process X is predictable if and only if $P(S < \infty, \Delta X_S \neq 0) = 0$ for every totally inaccessible stopping time S and the random variable $X_T \mathbb{1}_{\{T < \infty\}}$ is $\mathscr{F}_{T-}$-measurable for every predictable stopping time T.

PROOF.– The necessity follows from theorem 1.18 and proposition 1.11, so we prove the sufficiency. The second part of the assumption implies, in particular, that X is adapted. Let $\{T_n\}$ be a sequence of stopping times, exhausting jumps of X. By corollary 1.5, all T_n are accessible due to the first part of the assumptions. Using the definition of accessible stopping times and taking into account remark 1.6, we can find a sequence $\{S_n\}$ of predictable stopping times such that

$$\{\Delta X \neq 0\} \subseteq \bigcup_n [\![S_n]\!].$$

Replacing S_n by stopping times with the graphs $[\![S_n]\!] \setminus \left(\cup_{m < n} [\![S_m]\!] \right)$, we may assume that the graphs $[\![S_n]\!]$ are pairwise disjoint. Then we have

$$X = X_- + \sum_{n=1}^{\infty} \Delta X_{S_n} \mathbb{1}_{\{S_n < \infty\}} \mathbb{1}_{[\![S_n]\!]}.$$

Since $X_{S_n} \mathbb{1}_{\{S_n < \infty\}}$ is $\mathscr{F}_{S_n-}$-measurable by the second part of the assumption and $X_{S_n-} \mathbb{1}_{\{S_n < \infty\}}$ is $\mathscr{F}_{S_n-}$-measurable by proposition 1.11, the stochastic process $\Delta X_{S_n} \mathbb{1}_{\{S_n < \infty\}} \mathbb{1}_{[\![S_n]\!]}$ is predictable by lemma 1.6 for every n. Therefore, the process X is predictable. $\square$

At the end of this section, we consider an example from the general theory of stochastic processes in a series of exercises.

We start with a probability space $(\Omega, \mathscr{F}^0, \mathsf{P})$ and a random variable S with values in $\mathbb{R}_+$ on it. For example, we can take $\Omega = \mathbb{R}_+$, $\mathscr{F}^0 = \mathscr{B}(\mathbb{R}_+)$, $S(\omega) \equiv \omega$. For simplicity, we will assume that $\mathsf{P}(S = 0) = 0$.

Denote by $\mathscr{F}$ the completion of the σ-algebra $\mathscr{F}^0$ with respect to P. The continuation of P onto $\mathscr{F}$ is denoted by the same letter.

Now define the σ-algebra $\mathscr{F}_t^0$, $t \in \mathbb{R}_+$, as the smallest σ-algebra, with respect to which the random variable $S \wedge t$ is measurable. In other words,

$$\mathscr{F}_t^0 = \{S^{-1}(B) \colon B \in \mathscr{B}_t^0\},$$

where $\mathscr{B}_t^0$ is the σ-algebra on $\mathbb{R}_+$, generated by Borel subsets in $[0, t[$ and the atom $[t, \infty[$.

EXERCISE 1.27.– Show that $\mathbb{F}^0 := (\mathscr{F}_t^0)_{t \in \mathbb{R}_+}$ is a filtration.

EXERCISE 1.28.– Show that $\mathscr{F}_{t-}^0 = \mathscr{F}_t^0$ for every $t \in \mathbb{R}_+$.

EXERCISE 1.29.– Show that, for every $t \in \mathbb{R}_+$,

$$\mathscr{F}_{t+}^0 = \{S^{-1}(B) \colon B \in \mathscr{B}_{t+}^0\},$$

where the σ-algebra $\mathscr{B}_{t+}^0$ is obtained from the σ-algebra $\mathscr{B}_t^0$ by splitting the atom $[t, \infty[$ into two atoms, $\{t\}$ and $]t, \infty[$. Conclude that the filtration $\mathbb{F}^0$ is not right-continuous.

EXERCISE 1.30.– Show that S is a stopping time with respect to a filtration $(\mathscr{G}_t)_{t \in \mathbb{R}_+}$ if and only if $\mathscr{F}_{t+}^0 \subseteq \mathscr{G}_t$ for every $t \in \mathbb{R}_+$. In particular, S is a stopping time with respect to the filtration $(\mathbb{F}^0)^+ = (\mathscr{F}_{t+}^0)_{t \in \mathbb{R}_+}$, but is not a stopping time relative to $\mathbb{F}^0$.

Define a stochastic process $X = (X_t)_{t \in \mathbb{R}_+}$ by

$$X_t(\omega) = \mathbb{1}_{\{t \geqslant S(\omega)\}}.$$

It is clear that X is a càdlàg process. It is adapted with respect to some filtration if and only if S is a stopping time with respect to this filtration. So the statement in the previous exercise can be reformulated as follows: $(\mathbb{F}^0)^+$ is the smallest filtration, with respect to which X is adapted.

Let

$$\mathcal{N} = \{A \in \mathscr{F}: \mathsf{P}(A) = 0\}.$$

Put, for every $t \in \mathbb{R}_+$

$$\mathscr{F}_t := \{A \triangle B: A \in \mathscr{F}_{t+}^0, B \in \mathcal{N}\}.$$

According to exercises 1.2–1.4, the stochastic basis $\mathbb{B} := (\Omega, \mathscr{F}, \mathbb{F} = (\mathscr{F}_t)_{t \in \mathbb{R}_+}, \mathsf{P})$ satisfies the usual conditions. Moreover, $\mathbb{F}$ is the smallest (for given Ω, $\mathscr{F}$, and P) filtration such that the corresponding stochastic basis satisfies the usual conditions and X is adapted.

All subsequent statements refer to the stochastic basis $\mathbb{B}$. The first one says that the σ-algebra $\mathscr{F}_S$ coincides with the σ-algebra generated by S as a random variable up to a P-null set.

EXERCISE 1.31.– Prove that

$$\sigma\{S\} \vee \mathcal{N} = \mathscr{F}_{S-} = \mathscr{F}_S = \mathscr{F}_\infty.$$

In the next exercise, we give a simple characterization of all stopping times.

EXERCISE 1.32.– Show that a mapping $T: \Omega \to [0, \infty]$ is a stopping time (with respect to the filtration $\mathbb{F}$) if and only if it is $\mathscr{F}_\infty$-measurable and there is a constant $r \in [0, \infty]$ such that, P-a.s.,

$$T \geqslant S \quad \text{on} \quad \{S \leqslant r\}$$

and

$$T = r \quad \text{on} \quad \{S > r\}.$$

HINTS TO "ONLY IF".– If $\mathsf{P}(T < S) > 0$, then the set

$$U := \{u \in \mathbb{R}_+: \mathsf{P}(T \leqslant u < S) > 0\}$$

is not empty. For $u \in U$, deduce from the condition $\{T \leqslant u\} \in \mathscr{F}_u$ that $\{T \leqslant u\} \supseteq \{S > u\}$ a.s., i.e. $\mathsf{P}(S > u, T > u) = 0$. Conclude that $\mathsf{P}(T \leqslant u < S)$ does not increase in u on the set U and that $\inf U \in U$. Now take $\inf U$ as r.

To solve the next exercises use exercise 1.32 and the fact that a predictable stopping times is foretellable (theorem 1.13).

EXERCISE 1.33.– Assume that the random variable S has a discrete distribution. Prove that S is an accessible stopping time. Moreover, if S does not equal a.s. a constant, then S is not a predictable stopping time, and its graph is not represented as a countable union of the graphs of predictable stopping times up to an evanescent set.

EXERCISE 1.34.– Assume that the random variable S has a continuous distribution. Prove that S is a totally inaccessible stopping time. Show that the optional set $A :=$ $]\!]0, S]\!]$ does not admit a full section (see the remark after corollary 1.4).

EXERCISE 1.35.– Assume that the distribution of the random variable S has both continuous and discrete components. Show that S is neither an accessible, nor a totally inaccessible stopping time. Find the accessible and totally inaccessible parts of S.

1.6. Optional and predictable projections

Optional and *predictable projections* are a very useful and important tool in stochastic calculus. However, in this book, we will use only predictable projections and just to prove theorem 2.16, so this section may be skipped at first reading. Also, note that here we use some results from the theory of martingales developed in section 2.1.

Let a stochastic basis $\mathbb{B} = (\Omega, \mathscr{F}, \mathbb{F} = (\mathscr{F}_t)_{t \in \mathbb{R}_+}, \mathsf{P})$ satisfying the usual conditions be given.

THEOREM 1.20.– Let X be a bounded measurable process. There exists a unique (up to indistinguishability) bounded optional process Y such that

$$\mathsf{E} X_T \mathbb{1}_{\{T<\infty\}} = \mathsf{E} Y_T \mathbb{1}_{\{T<\infty\}} \qquad [1.4]$$

for every stopping time T. This process is called the optional projection of X and denoted by OX.

THEOREM 1.21.– Let X be a bounded measurable process. There exists a unique (up to indistinguishability) bounded predictable process Z such that

$$\mathsf{E} X_T \mathbb{1}_{\{T<\infty\}} = \mathsf{E} Z_T \mathbb{1}_{\{T<\infty\}} \qquad [1.5]$$

for every predictable stopping time T. This process is called the predictable projection of X and denoted by ΠX.

PROOF OF THEOREM 1.21.– Note that if two bounded measurable processes X^1 and X^2 satisfy $X^1 \leqslant X^2$ (up to an evanescent set) and have predictable projections ΠX^1 and ΠX^2, then $\Pi X^1 \leqslant \Pi X^2$ (up to an evanescent set). Indeed, if the predictable set $B := \{\Pi X^1 > \Pi X^2\}$ is not evanescent, then, by theorem 1.14 on predictable sections, there is a predictable stopping time T such that $\mathsf{P}(T < \infty) > 0$ and $[\![T]\!] \subseteq B$. Obviously, this contradicts [1.5].

In particular, a predictable projection is unique up to indistinguishability if it exists.

Let $\mathscr{H}$ be the class of bounded measurable processes whose predictable projection exists. It is clear that $\mathscr{H}$ is a linear space and contains constants. If $0 \leqslant X^1 \leqslant \ldots \leqslant X^n \leqslant \ldots \leqslant C$, then, as we have proved, $0 \leqslant \Pi X^1 \leqslant \ldots \leqslant \Pi X^n \leqslant \ldots \leqslant C$ everywhere, except on an evanescent set N. It follows from the theorem on monotone convergence that $\mathbb{1}_{(\Omega \times \mathbb{R}_+) \setminus N} \lim_n \Pi X^n$ is a predictable projection of the process $\lim_n X^n$. Therefore, by theorem A.3 on monotone classes, it is sufficient to prove the existence of predictable projections for processes

$$X = \xi \mathbb{1}_{[0,s]},$$

where ξ is a bounded random variable and $s \in \mathbb{R}_+$. Denote by M a bounded martingale such that $M_t = \mathsf{E}(\xi | \mathscr{F}_t)$ a.s. for every $t \in \mathbb{R}_+$, and put $Z := M_- \mathbb{1}_{[0,s]}$. Then, using exercise 2.14 and theorem 2.4, we get

$$\mathsf{E} X_T \mathbb{1}_{\{T < \infty\}} = \mathsf{E} \xi \mathbb{1}_{\{T \leqslant s\}} = \mathsf{E} M_\infty \mathbb{1}_{\{T \leqslant s\}} = \mathsf{E} \mathsf{E}(M_\infty | \mathscr{F}_{T-}) \mathbb{1}_{\{T \leqslant s\}}$$
$$= \mathsf{E} M_{T-} \mathbb{1}_{\{T \leqslant s\}} = \mathsf{E} Z_T \mathbb{1}_{\{T < \infty\}}$$

for every predictable stopping time T. Hence, $Z = \Pi X$. $\qquad \square$

PROOF OF THEOREM 1.20.– Completely similar to the proof of theorem 1.21, the only differences are that we should use the theorem on optional sections (corollary 1.4) instead of theorem 1.14 at the beginning of the proof, and we should take $Y := M \mathbb{1}_{[0,s]}$ at the end. $\qquad \square$

EXERCISE 1.36.– Let X be a bounded measurable process. Show that there exist a sequence $\{T_n\}$ of stopping times and an evanescent set N such that

$$\{OX \neq \Pi X\} \subseteq \left(\bigcup_n [\![T_n]\!] \right) \cup N.$$

HINT.– Use theorem A.3 on monotone classes.

The following theorem establishes a connection between projections and conditional expectations.

THEOREM 1.22.– Let X be a bounded measurable process. Then, P-a.s.,

$$(OX)_T \mathbb{1}_{\{T<\infty\}} = \mathsf{E}(X_T \mathbb{1}_{\{T<\infty\}}|\mathscr{F}_T) \qquad [1.6]$$

for every stopping time T and

$$(\Pi X)_T \mathbb{1}_{\{T<\infty\}} = \mathsf{E}(X_T \mathbb{1}_{\{T<\infty\}}|\mathscr{F}_{T-}) \qquad [1.7]$$

for every predictable stopping time T.

In particular, OX is an optional modification of a (defined up to a modification) stochastic process $\mathsf{E}(X_t|\mathscr{F}_t)$, and ΠX is a predictable modification of a stochastic process $\mathsf{E}(X_t|\mathscr{F}_{t-})$.

PROOF.– Let T be a stopping time and $B \in \mathscr{F}_T$. Then, T_B is a stopping time by proposition 1.5, and the equality [1.4], applied to T_B, yields

$$\mathsf{E}X_T \mathbb{1}_B \mathbb{1}_{\{T<\infty\}} = \mathsf{E}(OX)_T \mathbb{1}_B \mathbb{1}_{\{T<\infty\}}.$$

This equality is equivalent to [1.6] because the random variable $(OX)_T \mathbb{1}_{\{T<\infty\}}$ is $\mathscr{F}_T$-measurable by theorem 1.4. Similarly, let T be a predictable stopping time and $B \in \mathscr{F}_{T-}$. Then T_B is a predictable stopping time by proposition 1.14, and the equality [1.5] applied to T_B, yields

$$\mathsf{E}X_T \mathbb{1}_B \mathbb{1}_{\{T<\infty\}} = \mathsf{E}(\Pi X)_T \mathbb{1}_B \mathbb{1}_{\{T<\infty\}}.$$

This equality is equivalent to [1.7] because the random variable $(\Pi X)_T \mathbb{1}_{\{T<\infty\}}$ is $\mathscr{F}_{T-}$-measurable by proposition 1.11.   □

The assertions in the next exercise are simple consequences of section theorems (section 1.4).

EXERCISE 1.37.– Let X and Y be optional (respectively predictable) stochastic processes and $X_T = Y_T$ P-a.s. for every bounded (respectively bounded and predictable) stopping time. Prove that X and Y are indistinguishable.

REMARK 1.7.– The equalities [1.6] and [1.7] characterize the optional and predictable projections, respectively. Moreover, as exercise 1.37 shows, it is sufficient to check them only for *bounded* stopping times. On the contrary, it is necessary to check the equalities [1.4] and [1.5] for every stopping time, with finite or infinite values (see remark 1.5 to theorem 1.15). Indeed, let ξ be an $\mathscr{F}_\infty$-measurable bounded random variable with $\mathsf{E}\xi = 0$ (but $\mathsf{P}(\xi = 0) < 1$), $X = \xi \mathbb{1}_{[0,+\infty[}$. Then for every finite (or finite P-a.s.) stopping time T

$$\mathsf{E}X_T \mathbb{1}_{\{T<\infty\}} = \mathsf{E}X_T = \mathsf{E}\xi = 0.$$

However, $Y \equiv 0$ is neither an optional projection nor a predictable projection of X.

EXERCISE 1.38.– Justify the preceding assertion.

REMARK 1.8.– Optional and predictable projections can be defined for every nonnegative measurable process X as the limit in n of the corresponding projections of the processes $X \wedge n$. In this case, projections may take value $+\infty$. The equalities [1.4]–[1.7] are still valid.

EXERCISE 1.39.– Let X, Y and Z be bounded measurable stochastic processes. Moreover, assume that Y is optional and Z is predictable. Prove that

$$\Pi(OX) = \Pi X, \quad O(XY) = Y(OX), \quad \Pi(XZ) = Z(\Pi X). \qquad [1.8]$$

The following states that the difference between the progressively measurable and the optional σ-algebras is rather small. Recall that, however, these σ-algebras, in general, do not coincide.

EXERCISE 1.40.– Let X be a progressively measurable stochastic process. Prove that there exists an optional process Y such that $X_T = Y_T$ a.s. for every finite stopping time T. If X is the indicator of a set, then Y can be taken as an indicator function.

EXERCISE 1.41.– Let T be an accessible stopping time and $X = \mathbb{1}_{[T]}$. Show that, if a sequence $\{T_n\}$ of predictable stopping times embraces T, then

$$\{\Pi X > 0\} \subseteq \bigcup_n [\![T_n]\!]$$

up to an evanescent set, and there is an embracing sequence such that the equality holds up to an evanescent set.

Martingales and Processes with Finite Variation

2.1. Elements of the theory of martingales

This section outlines the basics of the theory of martingales with continuous time. Most results were obtained by Doob and are stated without proof. Unproved statements can be found in many textbooks on the theory of stochastic processes.

DEFINITION 2.1.– A stochastic process $X = (X_t)_{t \in \mathbb{R}_+}$ on a stochastic basis $\mathbb{B} = (\Omega, \mathscr{F}, \mathbb{F} = (\mathscr{F}_t)_{t \in \mathbb{R}_+}, \mathsf{P})$ is called a *martingale* (respectively a *submartingale*, respectively a *supermartingale*) if X is adapted, $\mathsf{E}|X_t| < \infty$ for all $t \in \mathbb{R}_+$ and

$$X_s = \mathsf{E}(X_t|\mathscr{F}_s) \quad (\text{respectively} \quad X_s \leqslant \mathsf{E}(X_t|\mathscr{F}_s),$$

$$\text{respectively} \quad X_s \geqslant \mathsf{E}(X_t|\mathscr{F}_s)) \quad \mathsf{P}\text{-a.s.}$$

for all $s, t \in \mathbb{R}_+$, $s < t$.

It is clear that X is a submartingale if and only if $-X$ is a supermartingale. Therefore, statements related to submartingales admit equivalent formulation for supermartingales and vice versa. We will usually formulate results only for supermartingales. It is also clear that a process is a martingale if and only if it is both a submartingale and a supermartingale.

EXERCISE 2.1.– Let a stochastic basis $\mathbb{B} = (\Omega, \mathscr{F}, \mathbb{F} = (\mathscr{F}_t)_{t \in \mathbb{R}_+}, \mathsf{P})$ and a probability Q on $(\Omega, \mathscr{F})$ be given. Denote by P_t and Q_t, the restrictions of P and Q, respectively onto the σ-algebra $\mathscr{F}_t$. Let $\mathsf{Q}_t = \mathsf{Q}_t^c + \mathsf{Q}_t^s$ be the Lebesgue decomposition of Q_t into the absolutely continuous component Q_t^c and the singular component Q_t^s relative to P_t. Put $Z_t = d\mathsf{Q}_t^c/d\mathsf{P}_t$, $Z = (Z_t)_{t \in \mathbb{R}_+}$. Show that Z is a supermartingale. Show that, if $\mathsf{Q}_t \ll \mathsf{P}_t$ for all $t \in \mathbb{R}_+$, then Z is a martingale.

PROPOSITION 2.1.– Let $X = (X_t)_{t \in \mathbb{R}_+}$ be a martingale, $f \colon \mathbb{R} \to \mathbb{R}$ be a convex function and $\mathsf{E}|f(X_t)| < \infty$ for all $t \in \mathbb{R}_+$. Then $f(X) = (f(X_t))_{t \in \mathbb{R}_+}$ is a submartingale. In particular, if $p \geqslant 1$ and $\mathsf{E}|X_t|^p < \infty$ for all $t \in \mathbb{R}_+$, then $(|X_t|^p)_{t \in \mathbb{R}_+}$ is a submartingale.

PROPOSITION 2.2.– Let $X = (X_t)_{t \in \mathbb{R}_+}$ be a submartingale, $f \colon \mathbb{R} \to \mathbb{R}$ a convex increasing function, and $\mathsf{E}|f(X_t)| < \infty$ for all $t \in \mathbb{R}_+$. Then $f(X) = (f(X_t))_{t \in \mathbb{R}_+}$ is a submartingale.

PROOF OF PROPOSITIONS 2.1 AND 2.2.– A real-valued convex function on the real line is continuous and, hence, measurable. Thus, the process $f(X)$ is adapted. The integrability of random variables $f(X_t)$ is due to the assumptions. Applying Jensen's inequality for conditional expectations, we get, for all $s < t$,

$$\mathsf{E}\big[f(X_t)|\mathscr{F}_s\big] \geqslant f\big[\mathsf{E}(X_t|\mathscr{F}_s)\big] \geqslant f(X_s)$$

(under the assumptions of proposition 2.1 the second inequality becomes an equality).

$\square$

Processes with independent increments are also a source of examples of martingales, sub- and supermartingales.

DEFINITION 2.2.– A stochastic process $X = (X_t)_{t \in \mathbb{R}_+}$ given on a probability space $(\Omega, \mathscr{F}, \mathsf{P})$ is called a *process with independent increments* if, for every $n \in \mathbb{N}$ and all $t_0, t_1, \ldots, t_n$ such that $0 = t_0 < t_1 < \cdots < t_n$, the random variables $X_{t_0}, X_{t_1} - X_{t_0}$, $\ldots, X_{t_n} - X_{t_{n-1}}$ are independent.

DEFINITION 2.3.– A stochastic process $X = (X_t)_{t \in \mathbb{R}_+}$ given on a stochastic basis $\mathbb{B} = (\Omega, \mathscr{F}, \mathbb{F} = (\mathscr{F}_t)_{t \in \mathbb{R}_+}, \mathsf{P})$ is called a *process with independent increments on* $\mathbb{B}$ if it is adapted relative to $\mathbb{F}$ and, for every $s, t \in \mathbb{R}_+$, $s < t$, the random variable $X_t - X_s$ and the σ-algebra $\mathscr{F}_s$ are independent.

EXERCISE 2.2.– Let a stochastic process $X = (X_t)_{t \in \mathbb{R}_+}$ be a process with independent increments on a stochastic basis $\mathbb{B}$. Show that X is a process with independent increments in the sense of definition 2.2.

EXERCISE 2.3.– Let a stochastic process $X = (X_t)_{t \in \mathbb{R}_+}$ given on a probability space $(\Omega, \mathscr{F}, \mathsf{P})$ be a process with independent increments in the sense of definition 2.2. Put $\mathscr{F}_t = \sigma\{X_s, s \leqslant t\}$. Show that X is a process with independent increments on the stochastic basis $(\Omega, \mathscr{F}, \mathbb{F} = (\mathscr{F}_t)_{t \in \mathbb{R}_+}, \mathsf{P})$.

EXERCISE 2.4.– Let a stochastic process $X = (X_t)_{t \in \mathbb{R}_+}$ be a process with independent increments on a stochastic basis $\mathbb{B} = (\Omega, \mathscr{F}, \mathbb{F}, \mathsf{P})$. Show that X is a process with independent increments on the stochastic basis $\mathbb{B}^\mathsf{P} = (\Omega, \mathscr{F}^\mathsf{P}, \mathbb{F}^\mathsf{P}, \overline{\mathsf{P}})$ (see exercise 1.3 for the notation).

EXERCISE 2.5.– Let a stochastic process $X = (X_t)_{t\in\mathbb{R}_+}$ be a process with independent increments on a stochastic basis $\mathbb{B} = (\Omega, \mathscr{F}, \mathbb{F} = (\mathscr{F}_t)_{t\in\mathbb{R}_+}, \mathsf{P})$. Assume that X is right-continuous in probability, i.e. for every $t \in \mathbb{R}_+$,

$$\lim_{\delta\downarrow 0} \mathsf{P}(|X_{t+\delta} - X_t| > \varepsilon) = 0 \qquad \text{for every } \varepsilon > 0.$$

Show that X is a process with independent increments on the stochastic basis $\mathbb{B}^+ = (\Omega, \mathscr{F}, \mathbb{F}^+, \mathsf{P})$ (see exercise 1.2 for the notation).

Exercises 2.3–2.5 show that if a stochastic process is a process with independent increments in the sense of definition 2.2 and is right-continuous in probability, then it is a process with independent increments on some stochastic basis satisfying the usual conditions. In accordance with this, the phrase like "W is a Wiener process on a stochastic basis $\mathbb{B}$" will always mean in the following that W is a Wiener process and, at the same time, is a process with independent increments on the stochastic basis $\mathbb{B}$.

EXERCISE 2.6.– Assume that X is a process with independent increments on a stochastic basis $\mathbb{B}$. Find necessary and sufficient conditions for X to be a martingale (submartingale).

EXERCISE 2.7.– Let $W = (W_t)_{t\in\mathbb{R}_+}$ be a Wiener process on a stochastic basis $(\Omega, \mathscr{F}, \mathbb{F}, \mathsf{P})$. Put $X_t = \exp(W_t - t/2)$. Show that W and $X = (X_t)_{t\in\mathbb{R}_+}$ are martingales.

EXERCISE 2.8.– Let $W = (W_t)_{t\in\mathbb{R}_+}$ be a Wiener process on a probability space $(\Omega, \mathscr{F}, \mathsf{P})$. Give an example of a filtration $\mathbb{F}$ on $(\Omega, \mathscr{F})$ such that W is adapted to $\mathbb{F}$ but is not a process with independent increments on the stochastic basis $(\Omega, \mathscr{F}, \mathbb{F}, \mathsf{P})$.

The following exercise is part of the assertion of theorem 2.1 below.

EXERCISE 2.9.– Let X be a right-continuous supermartingale. Show that the function $t \rightsquigarrow \mathsf{E}X_t$ is right-continuous on $\mathbb{R}_+$.

HINT.– Deduce from the supermartingale property that, for a given $t \in \mathbb{R}_+$, the sequence of random variables $\{X_{t+1/n}^-\}$ is uniformly integrable. Apply Fatou's lemma to $\{X_{t+1/n}\}$.

Definition 2.1 and the previous exercises do not require that the stochastic basis satisfies the usual conditions. From now on, we will assume that the stochastic basis $\mathbb{B} = (\Omega, \mathscr{F}, \mathbb{F} = (\mathscr{F}_t)_{t\in\mathbb{R}_+}, \mathsf{P})$ satisfies the usual conditions.

Note that if a stochastic basis is complete, a modification of an adapted process is again an adapted process. Thus, a modification of a martingale (respectively a submartingale, respectively a supermartingale), is a martingale (respectively a submartingale, respectively a supermartingale).

THEOREM 2.1 (without proof).– A supermartingale X has a càdlàg modification if and only if the function $t \rightsquigarrow \mathsf{E}X_t$ is right-continuous on $\mathbb{R}_+$. In particular, every martingale has a càdlàg modification.

ADDITION TO DEFINITION 2.1.– From now on, if not stated otherwise, we consider only càdlàg martingales, sub- and supermartingales.

THEOREM 2.2 (on convergence a.s.; without proof).– Let X be a supermartingale and

$$\sup_{t\in\mathbb{R}_+} \mathsf{E}X_t^- < \infty. \qquad [2.1]$$

Then with probability one, there exists a limit $X_\infty = \lim_{t\to\infty} X_t$ and $\mathsf{E}|X_\infty| < \infty$.

REMARK 2.1.– Quite often (especially if X is a martingale), theorem 2.2 is formulated with the assumption

$$\sup_{t\in\mathbb{R}_+} \mathsf{E}|X_t| < \infty, \qquad [2.2]$$

which is more strong, at first sight, than [2.1]. In fact, if X is a supermartingale, [2.1] and [2.2] are equivalent:

$$\mathsf{E}|X_t| = \mathsf{E}X_t + 2\mathsf{E}X_t^- \leqslant \mathsf{E}X_0 + 2\sup_{t\in\mathbb{R}_+} \mathsf{E}X_t^-.$$

EXERCISE 2.10.– Prove the second assertion in theorem 2.2.

In the next theorem necessary and sufficient conditions on a supermartingale X are given, under which the supermartingale property remains valid for *all* stopping times.

DEFINITION 2.4.– A supermartingale $X = (X_t)_{t\in\mathbb{R}_+}$ is called a *closed supermartingale* if there exists a random variable ξ such that

$$\mathsf{E}|\xi| < \infty \quad \text{and} \quad X_t \geqslant \mathsf{E}(\xi|\mathscr{F}_t) \quad \text{for all } t \in \mathbb{R}_+. \qquad [2.3]$$

A submartingale $X = (X_t)_{t\in\mathbb{R}_+}$ is called a *closed submartingale* if $-X$ is a closed supermartingale.

REMARK 2.2.– If X is a closed supermartingale, then, obviously, condition [2.1] in theorem 2.2 is satisfied. Therefore, for every stopping time T, the random variable X_T is well defined up to an evanescent set.

THEOREM 2.3 (without proof).– Let X be a closed supermartingale. Then, for every stopping times S and T such that $S \leqslant T$ a.s., random variables X_S and X_T are integrable and

$$X_S \geqslant \mathsf{E}(X_T | \mathscr{F}_S) \quad \text{P-a.s.} \tag{2.4}$$

REMARK 2.3.– It is, indeed, necessary that a supermartingale is closed in order for [2.4] to be true: take $S = t, T = \infty, \xi = X_\infty$.

REMARK 2.4.– Let X be a martingale. It follows from theorem 2.3 that we have *equality* in [2.4] if (and only if) X is both a closed supermartingale and a closed submartingale.

EXERCISE 2.11.– Let X be a martingale from exercise 2.7. Show that X is a closed supermartingale but not a closed submartingale. Give an example of stopping times $S \leqslant T$ such that equality in [2.4] does not hold.

COROLLARY 2.1.– Let X be a supermartingale. Then, for every stopping times S and T such that $S \leqslant T \leqslant N$ a.s. for some $N \in \mathbb{N}$, the random variables X_S and X_T are integrable and

$$X_S \geqslant \mathsf{E}(X_T | \mathscr{F}_S) \quad \text{P-a.s.} \tag{2.5}$$

PROOF.– Let $Y_t := X_{t \wedge N}$. It is easy to see that $Y = (Y_t)_{t \in \mathbb{R}_+}$ is a closed supermartingale. It remains to note that $X_S = Y_S$ and $X_T = Y_T$ a.s. □

COROLLARY 2.2.– Let X be a supermartingale. Then, for every stopping time T, the stopped process X^T is a supermartingale.

PROOF.– That X^T is adapted follows, e.g., from theorem 1.4. Next, the random variables $X_t^T = X_{T \wedge t}$ are integrable by the previous corollary. Fix s and t, $0 \leqslant s < t$. Define a supermartingale Y as in the previous proof with $N = t$. Then

$$X_{T \wedge t} = Y_T = Y_T \mathbb{1}_{\{T > s\}} + Y_T \mathbb{1}_{\{T \leqslant s\}} = Y_{T \vee s} \mathbb{1}_{\{T > s\}} + Y_T \mathbb{1}_{\{T \leqslant s\}}.$$

The random variable $Y_T \mathbb{1}_{\{T \leqslant s\}}$ is $\mathscr{F}_s$-measurable, hence,

$$\begin{aligned}
\mathsf{E}(X_{T \wedge t} | \mathscr{F}_s) &= \mathsf{E}(Y_{T \vee s} \mathbb{1}_{\{T > s\}} | \mathscr{F}_s) + Y_T \mathbb{1}_{\{T \leqslant s\}} \\
&= \mathbb{1}_{\{T > s\}} \mathsf{E}(Y_{T \vee s} | \mathscr{F}_s) + Y_T \mathbb{1}_{\{T \leqslant s\}} \\
&\leqslant \mathbb{1}_{\{T > s\}} Y_s + Y_T \mathbb{1}_{\{T \leqslant s\}} = Y_{T \wedge s} = X_{T \wedge s},
\end{aligned}$$

where the inequality follows from theorem 2.3 applied to the process Y and stopping times s and $T \vee s$. □

COROLLARY 2.3.– Let X be a martingale. Then, for every stopping time T, the stopped process X^T is a martingale.

DEFINITION 2.5.– A stochastic process $X = (X_t)_{t \in \mathbb{R}_+}$ is called *uniformly integrable* if the family $(X_t)_{t \in \mathbb{R}_+}$ is uniformly integrable. A progressively measurable stochastic process $X = (X_t)_{t \in \mathbb{R}_+}$ *belongs to the class* (D) if the family (X_T), where T runs over the class of all a.s. finite stopping times, is uniformly integrable. A progressively measurable stochastic process $X = (X_t)_{t \in \mathbb{R}_+}$ *belongs to the class* (DL) if, for every $t \in \mathbb{R}_+$, the process X^t (i.e. the process X stopped at the deterministic time t) belongs to the class (D). In other words, a progressively measurable stochastic process $X = (X_t)_{t \in \mathbb{R}_+}$ belongs to the class (DL) if, for every $t \in \mathbb{R}_+$, the family $(X_{T \wedge t})$, where T runs over the class of all stopping times, is uniformly integrable.

Let us introduce the following notation:

$\overline{\mathcal{M}}$ is a class of all martingales,

$\mathcal{M}$ is a class of all uniformly integrable martingales.

By corollary 2.3, $\overline{\mathcal{M}}$ is stable under stopping. By proposition A.1, if $M \in \overline{\mathcal{M}}$ and $t \in \mathbb{R}_+$, then $M^t \in \mathcal{M}$.

EXERCISE 2.12.– Prove that a martingale is uniformly integrable if and only if it is both a closed supermartingale and a closed submartingale.

THEOREM 2.4.– Let $M = (M_t)_{t \in \mathbb{R}_+} \in \mathcal{M}$. Then random variables M_t converge a.s. and in L^1 to a random variable M_∞ as $t \to \infty$, and for every stopping time T,

$$M_T = \mathsf{E}(M_\infty | \mathscr{F}_T) \quad \text{P-a.s.} \tag{2.6}$$

If T is a predictable stopping time, then P-a.s.

$$\mathsf{E}(M_\infty | \mathscr{F}_{T-}) = \mathsf{E}(M_T | \mathscr{F}_{T-}) = M_{T-} \tag{2.7}$$

(recall that $M_{\infty-} = M_\infty$).

With regard to [2.7], let us note that if $M \in \mathcal{M}$ and T is not a predictable stopping time, then $\mathsf{E}(M_\infty | \mathscr{F}_{T-})$ and M_{T-} may not be connected. For example, it is possible that $\mathsf{E}M_{T-} \neq \mathsf{E}M_\infty$ and that M_{T-} is not even integrable. We give an example of the first opportunity in the next exercise, and an example of the second opportunity is postponed until section 2.5 (example 2.4).

EXERCISE 2.13.– Construct an example of a martingale $M \in \mathcal{M}$ with $M_0 = 0$ and a stopping time T such that $\mathsf{E}|M_{T-}| < \infty$ and $\mathsf{E}M_{T-} \neq 0$.

HINT.– Take a process $\pi_t - \lambda t$, where (π_t) is a Poisson process with intensity λ, and stop it at the moment of the first jump of the Poisson process.

PROOF OF THEOREM 2.4.– It follows from theorem A.4 that M satisfies the assumptions of Ttheorem 2.2. Hence, for almost all ω, there is a limit $\lim_{t \to \infty} M_t(\omega) = M_\infty(\omega)$ and $\mathsf{E}|M_\infty| < \infty$. By theorem A.7, for every sequence $\{t_n\}$ converging to ∞, the sequence $\{M_{t_n}\}$ converges to M_∞ in L^1. Therefore, M_t converges in L^1 to M_∞ as $t \to \infty$. Passing to the limit as $s \to \infty$ in the equality

$$\int_B M_t \, dP = \int_B M_s \, dP, \quad B \in \mathscr{F}_t, \quad t < s,$$

we get

$$\int_B M_t \, dP = \int_B M_\infty \, dP, \quad B \in \mathscr{F}_t, \qquad \text{i.e., } M_t = \mathsf{E}(M_\infty|\mathscr{F}_t) \text{ P -a.s.}$$

Hence, the assumptions of theorem 2.3 are satisfied for both M and $-M$, and [2.6] follows.

Now let T be a predictable stopping time. Then, by theorem 1.13, there is a sequence $\{S_n\}$ of stopping times, which is a foretelling sequence for T. It follows from [2.6] that, for every n, a.s.

$$\mathsf{E}(M_\infty|\mathscr{F}_{S_n}) = M_{S_n}.$$

Let us pass to the limit as $n \to \infty$ in this relation. By Lévy's theorem for martingales with discrete time, the left side a.s. converges to $\mathsf{E}\left(M_\infty \middle| \bigvee_n \mathscr{F}_{S_n}\right)$. Moreover, $\bigvee_n \mathscr{F}_{S_n} = \mathscr{F}_{T-}$ by the second statement in theorem 1.3 (1). On the other hand, we have $S_n < T$ and $S_n \to T$ on the set $\{T > 0\}$. Hence, on this set, the variables M_{S_n} converge to M_{T-} (almost surely on the set $\{T = \infty\}$). With regard to the set $\{T = 0\}$, we have $S_n = 0$ for all n on it, and, hence, $M_{S_n} \equiv M_0 = M_{0-}$. Thus, we have proved that the expressions on the left and on the right in [2.7] coincide. That they are equal to the middle term follows from [2.6]. □

So far we have proved that every uniformly integrable martingale M is represented in the form

$$M_t = \mathsf{E}(\xi|\mathscr{F}_t), \tag{2.8}$$

where $E|\xi| < \infty$: take M_∞ as ξ. Conversely, let ξ be an integrable random variable. Define (up to a modification) a stochastic process $M = (M_t)_{t \in \mathbb{R}_+}$ by [2.8]. It follows from the tower property of conditional expectations that M is a martingale in the sense of definition 2.1, while proposition A.1 implies the uniform integrability of M. By theorem 2.1, M has a càdlàg modification (which is unique up to indistinguishability by proposition 1.1). For this modification, the statement of theorem 2.4 is true, and $M_\infty = E(\xi|\mathscr{F}_\infty)$ P-a.s.

EXERCISE 2.14.– Prove the previous assertion.

The next statement follows from [2.6] and proposition A.1.

COROLLARY 2.4.– Every uniformly integrable martingale belongs to the class (D), and every martingale belongs to the class (DL).

In section 2.2, we construct an example of a uniformly integrable supermartingale that does not belong to the class (D).

COROLLARY 2.5.– The class $\mathscr{M}$ is stable under stopping.

Let us denote by $L^p(\mathscr{F}_\infty)$, $1 \leqslant p < \infty$, the space $L^p(\Omega, \mathscr{F}_\infty, \mathsf{P}|_{\mathscr{F}_\infty})$ of (equivalence classes of P-a.s. equal) $\mathscr{F}_\infty$-measurable random variables ξ with $E|\xi|^p < \infty$. Let us also identify indistinguishable processes in $\mathscr{M}$. By theorem 2.4, the mapping $M \rightsquigarrow M_\infty$ maps $\mathscr{M}$ into $L^1(\mathscr{F}_\infty)$. It follows from [2.6] and proposition 1.1 that this mapping is injective, and it is surjective by exercise 2.14. In other words, this mapping is an isomorphism of linear spaces $\mathscr{M}$ and $L^1(\mathscr{F}_\infty)$.

Define for $p \in [1, \infty)$

$$\mathscr{M}^p := \{ M \in \mathscr{M} : E|M_\infty|^p < \infty \}.$$

Here, as in the case of $\mathscr{M}$, we identify indistinguishable processes, i.e. we interpret elements of the space $\mathscr{M}^p$ as equivalence classes of indistinguishable stochastic processes. It is clear that $\mathscr{M}^1 = \mathscr{M}$ and that the mapping $\mathscr{M}^p \ni M \rightsquigarrow M_\infty$ is an isomorphism of linear spaces $\mathscr{M}^p$ and $L^p(\mathscr{F}_\infty)$. Thus, the relation

$$\|M\|_{\mathscr{M}^p} := \left(E|M_\infty|^p \right)^{1/p}$$

supplies $\mathscr{M}^p$ with a norm which makes it a Banach space isomorphic to $L^p(\mathscr{F}_\infty)$. It follows from [2.6] and Jensen's inequality for conditional expectations that $\mathscr{M}^p$ is stable under stopping.

Let us formulate one more theorem belonging to Doob. To simplify the writing, we introduce the following notation which will also be used later. Let X be a stochastic process. Put

$$X_t^* := \sup_{0 \leqslant s \leqslant t} |X_s|, \qquad X_\infty^* := \sup_{t \in \mathbb{R}_+} |X_t|.$$

EXERCISE 2.15.– Let X be an adapted càdlàg stochastic process. Show that $X^* = (X_t^*)_{t \in \mathbb{R}_+}$ is an adapted stochastic process with values in $\mathbb{R}_+$, all trajectories of which are right-continuous and nondecreasing, and $X_\infty^*(\omega) = \lim_{t \to \infty} X_t^*(\omega)$ for all ω.

THEOREM 2.5 (without proof).– Let X be a nonnegative closed submartingale and $p \in (1, \infty)$. Then

$$\mathsf{E}(X_\infty^*)^p \leqslant \left(\frac{p}{p-1}\right)^p \mathsf{E}X_\infty^p.$$

COROLLARY 2.6.– Let M be a martingale, $t \in \mathbb{R}_+$, $p \in (1, \infty)$. Then

$$\mathsf{E}(M_t^*)^p \leqslant \left(\frac{p}{p-1}\right)^p \mathsf{E}|M_t|^p.$$

and

$$\mathsf{E}(M_\infty^*)^p \leqslant \left(\frac{p}{p-1}\right)^p \sup_{t \in \mathbb{R}_+} \mathsf{E}|M_t|^p. \qquad [2.9]$$

PROOF.– To prove the first inequality, apply theorem 2.5 to the submartingale $|M|$ stopped at deterministic time t. The second inequality follows from the first inequality. □

REMARK 2.5.– If $M \in \mathscr{M}$, then we can apply theorem 2.5 directly to a submartingale $|M|$ which is itself closed in this case, and obtain

$$\mathsf{E}(M_\infty^*)^p \leqslant \left(\frac{p}{p-1}\right)^p \mathsf{E}|M_\infty|^p$$

for every $p > 1$. The same inequality under the same assumptions is easy to deduce from [2.9], because it follows from [2.6] and Jensen's inequality for conditional expectations that $\mathsf{E}|M_t|^p \leqslant \mathsf{E}|M_\infty|^p$ for every $t \in \mathbb{R}_+$. But if M is a martingale satisfying only the assumptions of theorem 2.2, the above inequality may not be true.

EXERCISE 2.16.– Let M be the process X in exercises 2.7 and 2.11. Prove that $\mathsf{E}M_\infty^* = +\infty$ and, for every $p \in (1, \infty)$, the left side of [2.9] equals $+\infty$, while $\mathsf{E}|M_\infty|^p = 0$.

It is useful to introduce one more family of martingale spaces and reformulate some of previous results in new terms.

For $M \in \overline{\mathscr{M}}$ and $p \in [1, \infty)$, put

$$\|M\|_{\mathscr{H}^p} := \left(\mathsf{E}(M_\infty^*)^p\right)^{1/p}, \qquad \mathscr{H}^p := \{M \in \overline{\mathscr{M}} : \|M\|_{\mathscr{H}^p} < \infty\}.$$

Again, let us identify indistinguishable stochastic processes in $\mathscr{H}^p$. It is easy to check that $\mathscr{H}^p$ is a normed linear space. It is clear that $\mathscr{H}^p$ are stable under stopping, $\mathscr{H}^q \subseteq \mathscr{H}^p \subseteq \mathscr{M}^p \subseteq \mathscr{M}$ for $1 \leqslant p < q < \infty$.

COROLLARY 2.7.– Let M be a martingale and $p \in (1, \infty)$. The following statements are equivalent:

1) $\sup_{t \in \mathbb{R}_+} \mathsf{E}|M_t|^p < \infty$;

2) the family $\{|M_t|^p\}_{t \in \mathbb{R}_+}$ is uniformly integrable;

3) $M \in \mathscr{M}^p$;

4) $M \in \mathscr{H}^p$.

Moreover, in this case,

$$\|M\|_{\mathscr{M}^p} \leqslant \|M\|_{\mathscr{H}^p} \leqslant \frac{p}{p-1}\|M\|_{\mathscr{M}^p}. \qquad [2.10]$$

PROOF.– Implication $(4)\Rightarrow(3)$ has been mentioned above as obvious. If (3) holds, then [2.6] and Jensen's inequality for conditional expectations imply $|M_t|^p \leqslant \mathsf{E}(|M_\infty|^p|\mathscr{F}_t)$ for every $t \in \mathbb{R}_+$, and (2) follows from proposition A.1. Implication $(2)\Rightarrow(1)$ follows from theorem A.4. Implication $(1)\Rightarrow(4)$ is proved in corollary 2.6. The same corollary combined with remark 2.5 implies the second inequality in [2.10], while the first inequality is evident.   □

If $p = 1$, implications $(4)\Rightarrow(3)\Leftrightarrow(2)\Rightarrow(1)$ in corollary 2.7 are clearly still valid. The martingale X in exercises 2.7 and 2.11 satisfies (1) but does not satisfy the other three statements. The inclusion $\mathscr{H}^1 \subseteq \mathscr{M}$ is, in general, strict; see example 2.4.

It follows from the completeness of the space $\mathscr{M}^p$ and inequality [2.10] that:

COROLLARY 2.8.– $\mathscr{H}^p$ is a Banach space for $p \in (1, \infty)$.

The previous statement is true for $p = 1$ as well.

THEOREM 2.6.– $\mathscr{H}^1$ is a Banach space.

PROOF.– Let $\{M^n\}$ be a Cauchy sequence in $\mathscr{H}^1$. It is enough to find a convergent (in $\mathscr{H}^1$) subsequence $\{M^{n_k}\}$.

Choose n_k for $k = 1, 2, \dots$ so that $n_1 < n_2 < \dots < n_k < \dots$ and $\|M^n - M^m\|_{\mathscr{H}^1} \leqslant 2^{-k}$ for $n, m \geqslant n_k$. Let $\xi_k := (M^{n_{k+1}} - M^{n_k})_\infty^*$, then $\mathsf{E}\xi_k \leqslant 2^{-k}$. By the monotone convergence theorem,

$$\mathsf{E}\sum_{k=1}^{\infty} \xi_k = \sum_{k=1}^{\infty} \mathsf{E}\xi_k < \infty.$$

Therefore, the series $\sum_{k=1}^{\infty} \xi_k(\omega)$ converges for almost all ω. For these ω, the series

$$\sum_{k=1}^{\infty} (M_t^{n_{k+1}}(\omega) - M_t^{n_k}(\omega))$$

converges uniformly in $t \in \mathbb{R}_+$ by the Weierstrass test, therefore, $M_t^{n_k}(\omega)$ converges uniformly in $t \in \mathbb{R}_+$. Denote the limit of $M_t^{n_k}(\omega)$ by $M_t(\omega)$. For ω such that the series $\sum_{k=1}^{\infty} \xi_k(\omega)$ diverges, put $M_t(\omega) = 0$ for all $t \in \mathbb{R}_+$. Since $M_t = \lim_{k\to\infty} M_t^{n_k}$ a.s., M_t is $\mathscr{F}_t$-measurable, i.e. $M = (M_t)_{t \in \mathbb{R}_+}$ is an adapted stochastic process. Since the right-continuity and the existence of left limits preserve under the uniform convergence, M is a càdlàg process. Next, obviously,

$$(M - M^{n_k})_\infty^* \leqslant \sum_{i=k}^{\infty} \xi_i \quad \text{a.s.,}$$

hence,

$$\mathsf{E}M_\infty^* < \infty \qquad \text{and} \qquad \mathsf{E}(M - M^{n_k})_\infty^* \to 0.$$

It remains to note that it follows from the previous relation that, for every $t \in \mathbb{R}_+$, random variables $M_t^{n_k}$ converge to M_t in L^1. Therefore, we can pass to the limit as $k \to \infty$ in both sides of

$$\int_B M_s^{n_k} \, d\mathsf{P} = \int_B M_t^{n_k} \, d\mathsf{P}, \qquad s < t, \quad B \in \mathscr{F}_s,$$

and to obtain

$$\int_B M_s \, d\mathsf{P} = \int_B M_t \, d\mathsf{P}, \qquad s < t, \quad B \in \mathscr{F}_s,$$

Thus, M is a martingale. $\qquad\qquad\qquad\qquad\qquad\qquad\qquad\qquad\qquad\qquad\qquad\square$

REMARK 2.6.– We can prove directly that $\mathscr{H}^p$, $p > 1$, is complete. Indeed, let $\{M^n\}$ be a Cauchy sequence in $\mathscr{H}^p$, then it is a Cauchy sequence in $\mathscr{H}^1$ as well. We have just proved that there is a subsequence $\{M^{n_k}\}$ which converges in $\mathscr{H}^1$ to a martingale $M \in \mathscr{H}^1$. Moreover, $(M - M^{n_k})_\infty^* \to 0$ a.s. By Fatou's lemma,

$$\mathsf{E}\big(M_\infty^*\big)^p \leqslant \liminf_k \mathsf{E}\big((M^{n_k})_\infty^*\big)^p < \infty,$$

hence, $M \in \mathscr{H}^p$. For a given $\varepsilon > 0$, let k be such that $\|M^{n_l} - M^{n_k}\|_{\mathscr{H}^p} \leqslant \varepsilon$ if $l > k$. By Fatou's lemma,

$$\mathsf{E}\big((M - M^{n_k})_\infty^*\big)^p \leqslant \liminf_l \mathsf{E}\big((M^{n_l} - M^{n_k})_\infty^*\big)^p \leqslant \varepsilon^p.$$

We can introduce another useful norm on the space $\mathscr{H}^p$, $p \geqslant 1$, which is equivalent to $\| \cdot \|_{\mathscr{H}^p}$, and which is based on the Burkholder–Davis–Gundy inequality.

2.2. Local martingales

Unless otherwise stated, we will assume that a stochastic basis $\mathbb{B} = (\Omega, \mathscr{F}, \mathbb{F} = (\mathscr{F}_t)_{t \in \mathbb{R}_+}, \mathsf{P})$ satisfying the usual conditions is given.

DEFINITION 2.6.– A sequence $\{T_n\}$ of stopping times is called a *localizing sequence* if $T_1(\omega) \leqslant \ldots \leqslant T_n(\omega) \leqslant \ldots$ for all ω and $\lim_{n \to \infty} T_n(\omega) = +\infty$ for almost all ω.

The following technical assertion will be used repeatedly further.

LEMMA 2.1.–

1) If $\{T_n\}$ and $\{T_n'\}$ are localizing sequences, then $\{T_n \wedge T_n'\}$ is a localizing sequence.

2) Let a localizing sequence $\{T_n\}$ and, for every n, localizing sequences $\{T_{n,p}\}_{p \in \mathbb{N}}$ be given. Then there exists a localizing sequence $\{S_n\}$ such that, for every n,

$$S_n \leqslant T_n \wedge T_{n,p_n}.$$

for some natural numbers p_n.

PROOF.– (1) is obvious. Let us prove (2).

For every n, choose p_n such that

$$\mathsf{P}(T_{n,p_n} < n) \leqslant 2^{-n}.$$

This is possible because

$$\lim_{p\to\infty} \mathsf{P}(T_{n,p} < n) = 0.$$

Now put, for every n,

$$S_n := T_n \wedge \Big(\inf_{m\geqslant n} T_{m,p_m} \Big).$$

By proposition 1.3, S_n is a stopping time. The monotonicity of S_n in n follows from the monotonicity of T_n and the definition of S_n. Next,

$$\mathsf{P}(S_n < T_n \wedge n) \leqslant \mathsf{P}\Big(\inf_{m\geqslant n} T_{m,p_m} < n \Big) \leqslant \sum_{m=n}^{\infty} \mathsf{P}(T_{m,p_m} < m)$$

$$\leqslant \sum_{m=n}^{\infty} 2^{-m} = 2^{-n+1}.$$

Thus,

$$\sum_{n=1}^{\infty} \mathsf{P}(S_n < T_n \wedge n) < \infty.$$

By the Borel–Cantelli lemma, for almost all ω,

$$S_n(\omega) \geqslant T_n(\omega) \wedge n \quad \text{for } n \geqslant n(\omega). \tag{2.11}$$

However, it follows from definition 2.6 that, for almost all ω,

$$\lim_{n\to\infty} T_n(\omega) \wedge n = +\infty. \tag{2.12}$$

Combining [2.11] and [2.12], we get that, for almost all ω,

$$\lim_{n\to\infty} S_n(\omega) = +\infty. \qquad \square$$

DEFINITION 2.7.– An adapted càdlàg process $M = (M_t)_{t\in\mathbb{R}_+}$ is called a *local martingale* if there exists a localizing sequence $\{T_n\}$ of stopping times such that, for every n, we have $M^{T_n} \in \mathcal{M}$, i.e. the stopped process M^{T_n} is a uniformly integrable martingale. The class of all local martingales is denoted by $\mathcal{M}_{\text{loc}}$.

If $M \in \mathscr{M}_{\text{loc}}$ and $\{T_n\}$ is a localizing sequence of stopping times such that $M^{T_n} \in \mathscr{M}$ for every n, then we will say that $\{T_n\}$ is a localizing sequence for $M \in \mathscr{M}_{\text{loc}}$. A similar terminology will also be used for other "local classes" introduced later.

Our definition of a local martingale M always implies that $\mathsf{E}|M_0| < \infty$. Sometimes in the literature, another definition of a local martingale is used, which admits an arbitrary $\mathscr{F}_0$-measurable random variable as the initial value M_0, namely, a local martingale is understood as an adapted process M such that $M - M_0$ is a local martingale in the sense of our definition. It is easy to see that both definitions coincide if $\mathsf{E}|M_0| < \infty$.

It is clear that $\overline{\mathscr{M}} \subseteq \mathscr{M}_{\text{loc}}$: take $T_n = n$ as a localizing sequence. It is also clear that the class $\mathscr{M}_{\text{loc}}$ is stable under stopping: this follows from the definition and from the fact that $\mathscr{M}$ is stable under stopping, see corollary 2.5. The class $\mathscr{M}_{\text{loc}}$ is a linear space, use lemma 2.1 (1) to prove that the sum of two local martingales is a local martingale.

Local martingales play a very significant role in the following, and a considerable part of the book is devoted to studying their properties. In the end of the book, we introduce the notion of a σ-martingale, which also generalizes the notion of a martingale and includes the notion of a local martingale. The importance of the notion of a σ-martingale became apparent in the second half of the 1990s in connection with the first fundamental theorem of asset pricing.

THEOREM 2.7.– Let M be an adapted càdlàg process. The following statements are equivalent:

1) there exists a localizing sequence $\{T_n\}$ such that $M^{T_n} \in \mathscr{H}^1$ for every n;

2) $M \in \mathscr{M}_{\text{loc}}$;

3) there exists a localizing sequence $\{T_n\}$ such that $M^{T_n} \in \overline{\mathscr{M}}$ for every n;

4) there exists a localizing sequence $\{T_n\}$ such that $M^{T_n} \in \mathscr{M}_{\text{loc}}$ for every n.

The assertion of the theorem can be written in a symbolic form as $\mathscr{H}^1_{\text{loc}} = \mathscr{M}_{\text{loc}} = \overline{\mathscr{M}}_{\text{loc}} = (\mathscr{M}_{\text{loc}})_{\text{loc}}$.

PROOF.– Since $\mathscr{H}^1 \subseteq \mathscr{M} \subseteq \overline{\mathscr{M}} \subseteq \mathscr{M}_{\text{loc}}$, implications (1)$\Rightarrow$(2)$\Rightarrow$(3)$\Rightarrow$(4) are obvious.

We will first prove (4)$\Rightarrow$(1) under the assumption $M \in \mathscr{M}$. Put $T_n := \inf\{t: |M_t| > n\}$. By proposition 1.10, T_n is a stopping time. It is evident that T_n increase, and the regularity of trajectories of M implies $\lim_n T_n = \infty$, i.e. $\{T_n\}$ is a localizing sequence. Finally, $M^{T_n} \in \mathscr{M}$ due to corollary 2.5, and

$$\left(M^{T_n}\right)^*_\infty = M^*_{T_n} \leqslant n + |M_{T_n}|\mathbb{1}_{\{T_n < \infty\}} \in L^1,$$

i.e. $M^{T_n} \in \mathscr{H}^1$.

Now implication (4)$\Rightarrow$(1) is proved by applying lemma 2.1 (2) two times. The argument is standard and will often be replaced by the words "by localization". Assume first that $M \in \mathscr{M}_{\mathrm{loc}}$, then $M^{T_n} \in \mathscr{M}$ for some localizing sequence $\{T_n\}$. We have just proved that, for every n, there exists a localizing sequence $\{T_{n,p}\}_{p \in \mathbb{N}}$ such that $(M^{T_n})^{T_{n,p}} \in \mathscr{H}^1$ for all p. By lemma 2.1, there exists a localizing sequence $\{S_n\}$ such that

$$S_n \leqslant T_n \wedge T_{n,p_n}$$

for every n with some p_n. Then $M^{S^n} = \left((M^{T_n})^{T_{n,p_n}} \right)^{S^n} \in \mathscr{H}^1$ because $\mathscr{H}^1$ is stable under stopping.

Finally, let (4) hold, i.e. $M^{T_n} \in \mathscr{M}_{\mathrm{loc}}$ for a localizing sequence $\{T_n\}$. It was proved in the previous paragraph that, for every n, there exists a localizing sequence $\{T_{n,p}\}_{p \in \mathbb{N}}$ such that $(M^{T_n})^{T_{n,p}} \in \mathscr{H}^1$ for all p. The rest of the proof is the same as in the previous case. $\qquad\square$

THEOREM 2.8.– Let $M \in \mathscr{M}_{\mathrm{loc}}$. It is necessary and sufficient for $M \in \mathscr{M}$ that M belongs to the class (D).

PROOF.– The necessity was mentioned in corollary 2.4, so we prove the sufficiency. Let a local martingale M belong to the class (D). In particular, the family $(M_t)_{t \in \mathbb{R}_+}$ is uniformly integrable and $\mathrm{E}|M_t| < \infty$ for all t. Hence, it is enough to check the martingale property for M. Thus, let $s < t$, $B \in \mathscr{F}_s$, and let $\{T_n\}$ be a localizing sequence for $M \in \mathscr{M}_{\mathrm{loc}}$. Then:

$$\int_B M_s^{T_n} \, d\mathrm{P} = \int_B M_t^{T_n} \, d\mathrm{P}. \tag{2.13}$$

Note that $\lim_{n \to \infty} M_s^{T_n} = \lim_{n \to \infty} M_{s \wedge T_n} = M_s$ a.s., and the sequences $\{M_s^{T_n}\}$ and $\{M_t^{T_n}\}$ are uniformly integrable because M belongs to the class (D). So we can pass to the limit as $n \to \infty$ under the integral signs in [2.13] and to obtain

$$\int_B M_s \, d\mathrm{P} = \int_B M_t \, d\mathrm{P}. \qquad\qquad\square$$

In the proof, we can clearly see where the uniform integrability of values of M at stopping times is used. In fact, there exist local martingales M that are uniformly integrable but do not belong to the class (D) and, hence, are not martingales. An example will be given soon.

It follows from the following statement that a nonnegative local martingale is a supermartingale. However, in contrast to the case of discrete time, see [SHI 99, Chapter II, p. 100–101], it may not be a martingale, as we will see in the example mentioned in the previous paragraph.

Let us say that an adapted càdlàg process $X = (X_t)_{t \in \mathbb{R}_+}$ is a *local supermartingale* if there exists a localizing sequence $\{T_n\}$ of stopping times such that, for every n, the stopped process X^{T_n} is a supermartingale.

THEOREM 2.9.– A nonnegative local supermartingale is a supermartingale.

PROOF.– Let $X \geqslant 0$ and let X^{T_n} be a supermartingale for every n, where $\{T_n\}$ is a localizing sequence. Then, for every $t \in \mathbb{R}_+$,

$$\mathsf{E}X_{t \wedge T_n} = \mathsf{E}X_t^{T_n} \leqslant \mathsf{E}X_0^{T_n} = \mathsf{E}X_0,$$

and, by Fatou's lemma,

$$\mathsf{E}X_t \leqslant \liminf_{n \to \infty} \mathsf{E}X_{t \wedge T_n} \leqslant \mathsf{E}X_0 < \infty.$$

We obtain that the random variables X_t are integrable for every t.

Let $a > 0$ and $0 \leqslant s < t$. Obviously,

$$\mathsf{E}\big(X_t^{T_n} \wedge a \big| \mathscr{F}_s\big) \leqslant \mathsf{E}\big(X_t^{T_n} \big| \mathscr{F}_s\big) \wedge a \leqslant X_s^{T_n} \wedge a,$$

which implies, for every $B \in \mathscr{F}_s$,

$$\int_B (X_{t \wedge T_n} \wedge a)\, d\mathsf{P} \leqslant \int_B (X_{s \wedge T_n} \wedge a)\, d\mathsf{P}.$$

By the dominated convergence theorem, we may pass to the limit as $n \to \infty$ under the integral sign in both sides of the previous inequality:

$$\int_B (X_t \wedge a)\, d\mathsf{P} \leqslant \int_B (X_s \wedge a)\, d\mathsf{P}.$$

It remains to pass to the limit under the integral sign in the last inequality as $a \to \infty$ (this is possible by the monotone or dominated convergence theorems):

$$\int_B X_t\, d\mathsf{P} \leqslant \int_B X_s\, d\mathsf{P}.$$

Thus, X is a supermartingale. □

COROLLARY 2.9.– Let M be a nonnegative local martingale with $M_0 = 0$. Then M is indistinguishable with the identically zero process.

PROOF.– By theorem 2.9, M is a supermartingale. Hence, for every $t \in \mathbb{R}_+$,

$$0 \leqslant \mathsf{E}M_t \leqslant \mathsf{E}M_0 = 0.$$

Since M_t is nonnegative, we have $M_t = 0$ a.s. It remains to use proposition 1.1. $\square$

We will often refer later to the next theorem which describes jumps of local martingales at predictable stopping times. The readers, not familiar with the definition of conditional expectation for nonintegrable random variables, are recommended to read section A.3 in the Appendix.

THEOREM 2.10.– Let $M \in \mathcal{M}_{\mathrm{loc}}$ and T be a predictable stopping time. Then a.s.

$$\mathsf{E}\big(|\Delta M_T|\mathbb{1}_{\{T<\infty\}}\big|\mathscr{F}_{T-}\big) < \infty \quad \text{and} \quad \mathsf{E}\big(\Delta M_T \mathbb{1}_{\{T<\infty\}}\big|\mathscr{F}_{T-}\big) = 0.$$

PROOF.– Let $\{T_n\}$ be a localizing sequence for $M \in \mathcal{M}_{\mathrm{loc}}$. Applying theorem 2.4 to the uniformly integrable martingale M^{T_n}, we get, for every n,

$$\mathsf{E}|M_{T \wedge T_n}| < \infty \quad \text{and} \quad \mathsf{E}\big(M_{T \wedge T_n}\big|\mathscr{F}_{T-}\big) = M_{(T \wedge T_n)-} \quad \text{a.s. on } \{T < \infty\}.$$

Since $\{T \leqslant T_n\} \in \mathscr{F}_{T-}$ by theorem 1.2 (5), we have $B_n := \{T \leqslant T_n\} \cap \{T < \infty\} \in \mathscr{F}_{T-}$, hence

$$\mathsf{E}|M_T|\mathbb{1}_{B_n} < \infty \quad \text{and} \quad \mathsf{E}\big(M_T\big|\mathscr{F}_{T-}\big) = M_{T-} \quad \text{a.s. on } B_n.$$

It remains to note that $\cup_n B_n = \{T < \infty\}$ a.s. and $\{T < \infty\} \in \mathscr{F}_{T-}$. $\square$

In the discrete time, a local martingale is an adapted process for which the one-step martingale property holds but the integrability may fail. It is a mistake to imagine that in the case of continuous time the situation is similar. The following two examples illustrate this point.

EXAMPLE 2.1.– Let $W = (W_t)_{t \in \mathbb{R}_+}$ be a standard Wiener process on a complete probability space $(\Omega, \mathscr{F}, \mathsf{P})$. Assume that a filtration $\mathbb{G} = (\mathscr{G}_t)_{t \in \mathbb{R}_+}$ on $(\Omega, \mathscr{F})$ is such that the stochastic basis $\mathbb{B}_{\mathbb{G}} := (\Omega, \mathscr{F}, \mathbb{G}, \mathsf{P})$ satisfies the usual conditions and W is a process with independent increments on $\mathbb{B}_{\mathbb{G}}$; see definition 2.3 and exercises 2.3–2.5). Recall that $X_t = \exp(W_t - t/2)$ is a martingale on $\mathbb{B}_{\mathbb{G}}$ according to exercise 2.7.

Now let us take a one-to-one increasing (and necessarily continuous) function

$$\psi\colon [0,1) \to [0,\infty).$$

Put $\mathscr{F}_t := \mathscr{G}_{\psi(t)}$ for $t \in [0, 1)$ and $\mathscr{F}_t := \mathscr{F}$ for $t \in [1, \infty)$, $\mathbb{F} := (\mathscr{F}_t)_{t \in \mathbb{R}_+}$. It is clear that $\mathbb{F}$ is a filtration and the stochastic basis $\mathbb{B}_{\mathbb{F}} := (\Omega, \mathscr{F}, \mathbb{F}, \mathsf{P})$ satisfies the usual conditions. Put also

$$\widetilde{M}_t := \begin{cases} X_{\psi(t)}, & \text{if } t \in [0, 1), \\ 0, & \text{if } t \in [1, \infty). \end{cases}$$

The process $\widetilde{M} = (\widetilde{M}_t)_{t \in \mathbb{R}_+}$ is adapted relative to $\mathbb{F}$, and its trajectories are continuous everywhere except possibly the point $t = 1$, where they are right-continuous. Moreover, it follows from the law of iterated logarithm for the Wiener process that

$$\lim_{t \to \infty} (W_t - t/2) = -\infty \quad \text{a.s.,}$$

hence the set $N := \{\omega \colon \limsup_{t \to \infty} X_t > 0\}$ has zero measure. Now define, for all t,

$$M_t(\omega) := \begin{cases} \widetilde{M}_t(\omega), & \text{if } \omega \notin N, \\ 0, & \text{if } \omega \in N. \end{cases}$$

The process $M = (M_t)_{t \in \mathbb{R}_+}$ is continuous, nonnegative, adapted relative to $\mathbb{F}$ (because the basis $\mathbb{B}_{\mathbb{F}}$ is complete). Next, for $0 \leqslant s < t < 1$,

$$\mathsf{E}M_t = \mathsf{E}X_{\psi(t)} = \mathsf{E}X_0 = 1$$

and

$$\mathsf{E}(M_t | \mathscr{F}_s) = \mathsf{E}(X_{\psi(t)} | \mathscr{F}_{\psi(s)}) = X_{\psi(s)} = M_s,$$

and, for $t \geqslant 1$,

$$M_t = 0.$$

Hence, M is a supermartingale on $\mathbb{B}_{\mathbb{F}}$, which has the martingale property on the intervals $[0, 1)$ and $[1, \infty)$, but this property is "broken" at $t = 1$. We assert that, additionally, M is a local martingale.

Indeed, put $T_n := \inf \{t \colon M_t > n\}$. By proposition 1.10 (1), T_n is a stopping time on $\mathbb{B}_{\mathbb{F}}$. It is obvious that T_n are increasing. Since all trajectories of M are bounded, for every ω, $T_n(\omega) = \infty$ for n large enough. Thus, $\{T_n\}$ is a localizing sequence.

Now we check that M^{T_n} is a uniformly integrable martingale on $\mathbb{B}_{\mathbb{F}}$ for every n. Since M^{T_n} is a supermartingale on $\mathbb{B}_{\mathbb{F}}$, see Corollary 2.2, it is enough to show that

$\mathsf{E}M_{T_n} = \mathsf{E}M_0 = 1$, see Exercise 2.17 below. Let $S_n := \inf\{t\colon X_t > n\}$. Then S_n is a stopping time on $\mathbb{B}_{\mathbb{G}}$ and clearly $\psi(T_n) = S_n$ a.s., where $\psi(\infty) := \infty$. Since $0 \leqslant X^{S_n} \leqslant n$, the stopped process X^{S_n}, being a martingale on $\mathbb{B}_{\mathbb{G}}$, is a uniformly integrable martingale on $\mathbb{B}_{\mathbb{G}}$. Hence,

$$\mathsf{E}X_{S_n} = \mathsf{E}X_0 = 1$$

and

$$\mathsf{E}M_{T_n} = \mathsf{E}X_{S_n} = 1.$$

EXERCISE 2.17.– Let a supermartingale X satisfy the assumptions of theorem 2.3. Show that $X \in \mathscr{M}$ if and only if $\mathsf{E}X_\infty = \mathsf{E}X_0$.

EXERCISE 2.18.– Using example 2.1, construct an example of a local martingale $M = (M_t)_{t\in\mathbb{R}_+}$, such that $M_0 = M_1 = 0$ and $\mathsf{P}(M_t = 0) = 0, 0 < t < 1$.

In example 2.1, $M = (M_t)_{t\in\mathbb{R}_+}$ is a nonnegative continuous local martingale, whose expectation $\mathsf{E}M_t$ as a function of t has a jump at $t = 1$. In the following classical example, we construct a nonnegative continuous local martingale $M = (M_t)_{t\in\mathbb{R}_+}$, whose expectation $\mathsf{E}M_t$ is continuous in $t \in [0, \infty]$ (which implies that M is uniformly integrable) but strictly decreasing in t. Thus, a supermartingale $M = (M_t)_{t\in\mathbb{R}_+}$ is uniformly integrable but does not belong to the class (D) and even to the class (DL).

EXERCISE 2.19.– Let $X = (X_t)_{t\in\mathbb{R}_+}$ be a nonnegative supermartingale. Assume that the function $t \rightsquigarrow \mathsf{E}X_t$ is continuous on $[0, \infty]$ (note that the random variable X_∞ is defined and integrable by theorem 2.2). Prove that the process X is uniformly integrable. The continuity of X is not assumed.

EXAMPLE 2.2.– Let $(\Omega, \mathscr{F}, \mathsf{P})$ be a complete probability space with a three-dimensional (3D) Brownian motion $W = (W^1, W^2, W^3)$ on it. This means that W is a stochastic process with values in $\mathbb{R}^3$, whose components $W^i = (W_t^i)_{t\in\mathbb{R}_+}$, $i = 1, 2, 3$, are independent standard Wiener processes. A filtration $\mathbb{F} = (\mathscr{F}_t)_{t\in\mathbb{R}_+}$ on $(\Omega, \mathscr{F})$ is chosen in such a way that the stochastic basis $\mathbb{B} := (\Omega, \mathscr{F}, \mathbb{F}, \mathsf{P})$ satisfies the usual conditions, W is adapted relative to $\mathbb{F}$, and, for any $0 \leqslant s < t$, the random vector $W_t - W_s$ and the σ-algebra $\mathscr{F}_s$ are independent. Put $X_t = W_t + (0, 0, 1)$. The process $M = (M_t)_{t\in\mathbb{R}_+}$ that we want to construct is given by

$$M_t = \frac{1}{\|X_t\|}, \quad t \in \mathbb{R}_+, \tag{2.14}$$

where $\|\cdot\|$ is the Euclidean norm in $\mathbb{R}^3$. However, we cannot do it directly because the denominator in [2.14] may vanish.

So let us first define

$$T_n := \inf \{t \colon \|X_t\| < 1/n\}.$$

By proposition 1.10, T_n is a stopping time for every n. Obviously, the sequence T_n increases in n. Define the process $M^n = (M_t^n)_{t \in \mathbb{R}_+}$ by

$$M_t^n = \frac{1}{\|X_t^{T_n}\|}, \quad t \in \mathbb{R}_+.$$

It is clear that M^n is a continuous adapted nonnegative stochastic process bounded from above by n. The key fact is that M^n is a local martingale and, hence, a uniformly integrable martingale. The proof of this fact is based on Itô's formula, see theorem 3.10 and exercise 3.9 in section 3.2, and what is essential is that the function $\mathbf{x} \rightsquigarrow 1/\|\mathbf{x}\|$ is harmonic on $\mathbb{R}^3 \setminus \{0\}$.

Since X is continuous, we have $M_{T_n}^n = n$ on the set $\{T_n < \infty\}$. Next, $\mathsf{E}M_{T_n}^n = \mathsf{E}M_0^n = 1$ by theorem 2.4, hence,

$$1 = \mathsf{E}M_{T_n}^n \geqslant \mathsf{E}M_{T_n}^n \mathbb{1}_{\{T_n < \infty\}} = n\mathsf{P}(T_n < \infty),$$

hence $\mathsf{P}(T_n < \infty) \leqslant 1/n$. Therefore, $\mathsf{P}(\cap_n \{T_n < \infty\}) = 0$ and, since T_n are increasing,

$$\lim_n T_n = \infty \quad \mathsf{P}\text{-a.s.}$$

Now we can define M by [2.14] for ω such that $\lim_n T_n(\omega) = \infty$; otherwise, put $M_t(\omega) = 1$ for all t. It is clear that M is a continuous adapted process. Since M^{T_n} and M^n are indistinguishable, we have $M^{T_n} \in \mathcal{M}$ for every n. Therefore, $M \in \mathcal{M}_{\text{loc}}$. In particular, M is a supermartingale by theorem 2.9.

Now let us compute $\mathsf{E}M_t$. The vector X_t has a Gaussian distribution with the density

$$f_t(\mathbf{x}) = \frac{1}{\sqrt{(2\pi t)^3}} e^{-\frac{1}{2t}\|\mathbf{x} - \mathbf{x}_0\|^2}, \quad \mathbf{x} = (x_1, x_2, x_3), \quad \mathbf{x}_0 = (0, 0, 1).$$

Hence,

$$\mathsf{E}M_t = \int\limits_{-\infty}^{+\infty} \int\limits_{-\infty}^{+\infty} \int\limits_{-\infty}^{+\infty} \frac{1}{\|\mathbf{x}\|} f_t(\mathbf{x}) \, dx_1 \, dx_2 \, dx_3.$$

Passing to the spherical coordinates

$$x_1 = r \sin \varphi \cos \vartheta, \quad x_2 = r \sin \varphi \sin \vartheta, \quad x_3 = r \cos \varphi,$$

$$0 \leqslant r < \infty, \quad 0 \leqslant \varphi \leqslant \pi, \quad 0 \leqslant \vartheta < 2\pi,$$

we get

$$
\begin{aligned}
\mathsf{E} M_t &= \frac{1}{\sqrt{(2\pi t)^3}} \int_{-\infty}^{+\infty} \int_{-\infty}^{+\infty} \int_{-\infty}^{+\infty} \frac{1}{\|\mathbf{x}\|} e^{-\frac{1}{2t}\|\mathbf{x}-\mathbf{x}_0\|^2} \, dx_1 \, dx_2 \, dx_3 \\
&= \frac{1}{\sqrt{(2\pi t)^3}} \int_0^{+\infty} \int_0^\pi \int_0^{2\pi} \frac{1}{r} e^{-\frac{1}{2t}\left[r^2 \sin^2 \varphi + (r\cos\varphi - 1)^2\right]} r^2 \sin \varphi \, d\vartheta \, d\varphi \, dr \\
&= \frac{2\pi}{\sqrt{(2\pi t)^3}} \int_0^{+\infty} r e^{-\frac{r^2+1}{2t}} \int_0^\pi e^{\frac{r\cos\varphi}{t}} \sin \varphi \, d\varphi \, dr \\
&= \frac{2\pi}{\sqrt{(2\pi t)^3}} \int_0^{+\infty} r e^{-\frac{r^2+1}{2t}} \int_{-1}^{1} e^{\frac{rz}{t}} \, dz \, dr \\
&= \frac{2\pi t}{\sqrt{(2\pi t)^3}} \int_0^{+\infty} e^{-\frac{r^2+1}{2t}} \left(e^{\frac{r}{t}} - e^{-\frac{r}{t}}\right) dr \\
&= \frac{1}{\sqrt{2\pi t}} \int_0^{+\infty} \left(e^{-\frac{(r-1)^2}{2t}} - e^{-\frac{(r+1)^2}{2t}}\right) dr \\
&= \frac{1}{\sqrt{2\pi}} \left(\int_{-1/\sqrt{t}}^{+\infty} e^{-\frac{y^2}{2}} \, dy - \int_{+1/\sqrt{t}}^{+\infty} e^{-\frac{y^2}{2}} \, dy \right) \\
&= \Phi(1/\sqrt{t}) - \Phi(-1/\sqrt{t}),
\end{aligned}
$$

where $\Phi(\cdot)$ is the distribution function of the standard normal law. Thus, the function $t \rightsquigarrow \mathsf{E} M_t$ is strictly decreasing and continuous on $[0, \infty)$, and $\lim_{t \uparrow \infty} \mathsf{E} M_t = 0$. We conclude that M is not a martingale. Note also that, by theorem 2.2, there exists a.s. a limit $\lim_{t \uparrow \infty} M_t = M_\infty$, and $\mathsf{E} M_\infty \leqslant 0$ by Fatou's lemma, that is, $M_\infty = 0$ a.s. By exercise 2.19, the process M is uniformly integrable.

We can also prove the uniform integrability in a direct way using the Vallée–Poussin criterion. Indeed, similarly to the previous computations,

$$
\begin{aligned}
\mathsf{E} M_t^2 &= \frac{1}{\sqrt{(2\pi t)^3}} \iiint\limits_{\mathbb{R}^3} \frac{1}{\|\mathbf{x}\|^2} e^{-\frac{1}{2t}\|\mathbf{x}-\mathbf{x}_0\|^2} \, dx_1 \, dx_2 \, dx_3 \\
&\leqslant \frac{2}{\sqrt{(2\pi t)^3}} \iiint\limits_{\|\mathbf{x}\|>1/2} \frac{1}{\|\mathbf{x}\|} e^{-\frac{1}{2t}\|\mathbf{x}-\mathbf{x}_0\|^2} \, dx_1 \, dx_2 \, dx_3 \\
&\quad + \frac{e^{-\frac{1}{8t}}}{\sqrt{(2\pi t)^3}} \iiint\limits_{\|\mathbf{x}\|\leqslant 1/2} \frac{1}{\|\mathbf{x}\|^2} \, dx_1 \, dx_2 \, dx_3 \\
&\leqslant 2 + \frac{e^{-\frac{1}{8t}}}{\sqrt{(2\pi t)^3}} \int\limits_0^{1/2} \int\limits_0^{\pi} \int\limits_0^{2\pi} \frac{1}{r^2} r^2 \sin\varphi \, d\vartheta \, d\varphi \, dr \\
&\leqslant 2 + 2\pi \sup_{t>0} \frac{e^{-\frac{1}{8t}}}{\sqrt{(2\pi t)^3}} < \infty,
\end{aligned}
$$

i.e.,

$$
\sup_{t\in\mathbb{R}_+} \mathsf{E} M_t^2 < \infty. \qquad\qquad [2.15]
$$

Thus, M is a continuous local martingale satisfying [2.15], however, $M \notin \mathscr{H}^2$, because M is not a martingale. So this example also shows that it is not enough to assume in corollary 2.7 that M is a local martingale.

EXERCISE 2.20.– Show that $\mathsf{P}(T_n < \infty) = 1/n$ in example 2.2.

2.3. Increasing processes and processes with finite variation

We will assume that a stochastic basis $\mathbb{B} = (\Omega, \mathscr{F}, \mathbb{F} = (\mathscr{F}_t)_{t\in\mathbb{R}_+}, \mathsf{P})$ satisfying the usual conditions is given.

DEFINITION 2.8.– An adapted stochastic process $A = (A_t)_{t\in\mathbb{R}_+}$ is called an *increasing process* if, for all ω, trajectories $t \rightsquigarrow A_t(\omega)$ are right-continuous, start from 0, i.e. $A_0(\omega) = 0$, and are nondecreasing, i.e. $A_s(\omega) \leqslant A_t(\omega)$ for every $s, t \in \mathbb{R}_+$, $s \leqslant t$. An adapted process $A = (A_t)_{t\in\mathbb{R}_+}$ is called a *process with finite variation* if, for all ω, trajectories $t \rightsquigarrow A_t(\omega)$ are right-continuous, start from 0, and have a finite variation on $[0, t]$ for every $t \in \mathbb{R}_+$. The class of all increasing processes is denoted by $\mathscr{V}^+$, and the class of all processes with finite variation is denoted by $\mathscr{V}$.

EXERCISE 2.21.– Let $A \in \mathcal{V}$. Show that the following conditions are equivalent:

1) there is a process $B \in \mathcal{V}^+$ indistinguishable from A;

2) for almost all ω, trajectories $t \rightsquigarrow A_t(\omega)$ are nondecreasing;

3) $A_s \leqslant A_t$ a.s. for all $s, t \in \mathbb{R}_+$, $s < t$.

Increasing processes and processes with finite variation relative to the filtration $\mathscr{F}_t \equiv \mathscr{F}$ will be called increasing processes in the wide sense and processes with finite variation in the wide sense, respectively. In other words, to define corresponding processes in the wide sense, we should omit the word "adapted" in definition 2.8.

Trajectories of any process with finite variation, besides being right-continuous, have finite left limits at every point in $(0, \infty)$. Thus, processes in $\mathcal{V}$ are optional.

If $A \in \mathcal{V}^+$, then, for all ω, there exists a limit $A_\infty(\omega) = \lim_{t \uparrow \infty} A_t(\omega) \in [0, +\infty]$.

It is clear that $\mathcal{V}^+$ is a convex cone, $\mathcal{V}$ is a linear space, $\mathcal{V}^+$ and $\mathcal{V}$ are stable under stopping. If a process A is such that $A^{T_n} \in \mathcal{V}^+$ (respectively $A^{T_n} \in \mathcal{V}$) for all n for some localizing sequence $\{T_n\}$, then A is indistinguishable with a process in $\mathcal{V}^+$ (respectively in $\mathcal{V}$). This explains why we do not introduce corresponding local classes.

Let $A \in \mathcal{V}$. The variation of a trajectory $s \rightsquigarrow A_s(\omega)$ on the interval $[0, t]$ is denoted by $\mathrm{Var}\,(A)_t(\omega)$. It follows from the next proposition that $\mathrm{Var}\,(A) = \big(\mathrm{Var}\,(A)_t\big)_{t \in \mathbb{R}_+}$ is a stochastic process (even if A is a process with finite variation in the wide sense only). All relations in this proposition are understood to hold for all trajectories.

PROPOSITION 2.3.– Let $A \in \mathcal{V}$. There is a unique pair (B, C) of processes $B, C \in \mathcal{V}^+$ such that

$$A = B - C, \qquad \mathrm{Var}\,(A) = B + C.$$

In particular, $\mathrm{Var}\,(A) \in \mathcal{V}^+$. If A is predictable, B, C and $\mathrm{Var}\,(A)$ are predictable.

PROOF.– Put

$$B = \frac{\mathrm{Var}\,(A) + A}{2}, \qquad C = \frac{\mathrm{Var}\,(A) - A}{2}, \qquad [2.16]$$

see [A.8]. The required pathwise properties of processes B, C and $\mathrm{Var}\,(A)$ follow from results given in Appendix A.4. Next, by [A.7], for every $t \in \mathbb{R}_+$,

$$\mathrm{Var}\,(A)_t = \lim_{n\to\infty} \sum_{k=1}^{2^n} |A_{kt2^{-n}} - A_{(k-1)t2^{-n}}|,$$

which implies that $\mathrm{Var}\,(A)$ is adapted. Hence, each of the processes B, C and $\mathrm{Var}\,(A)$ belongs to $\mathscr{V}^+$. Finally, let A be predictable. Since

$$\Delta\,\mathrm{Var}\,(A) = |\Delta A|,$$

processes $\Delta\,\mathrm{Var}\,(A)$ and $\mathrm{Var}\,(A)$ are predictable by proposition 1.8. Hence, B and C are also predictable in view of [2.16].   □

In particular, it follows from this proposition that every process in $\mathscr{V}$ can be represented as the difference of two increasing processes from $\mathscr{V}^+$. The converse is, of course, also true. Thus, $\mathscr{V} = \mathscr{V}^+ - \mathscr{V}^+$.

DEFINITION 2.9.– A process $A \in \mathscr{V}$ is called *purely discontinuous*, if, for all ω and $t \in \mathbb{R}_+$,

$$A_t(\omega) = \sum_{0 < s \leqslant t} \Delta A_s(\omega).$$

The class of all purely discontinuous processes with finite variation is denoted by $\mathscr{V}^d$.

In the next exercise, the equality is understood to hold for all trajectories.

EXERCISE 2.22.– Let $A \in \mathscr{V}$. Show that there are a continuous process $B \in \mathscr{V}$ and a purely discontinuous process $C \in \mathscr{V}^d$ such that $A = B + C$.

Trajectories of a process $A \in \mathscr{V}$ have finite variation on every finite interval (unlike, say, a Wiener process). Consequently, we can define an "integral process" $\int_0^t H_s\,dA_s$ as the pathwise Lebesgue–Stieltjes integral for a wide class of processes H.

Let A be a process with finite variation in the wide sense and H be a measurable stochastic process. Then, for every ω, the trajectory $t \rightsquigarrow H_t(\omega)$ is a measurable (with respect to the Borel σ-algebra $\mathscr{B}(\mathbb{R}_+)$) function. Therefore, we can define

$$Y_t(\omega) := \begin{cases} \int\limits_0^t H_s(\omega)\,dA_s(\omega), & \text{if } \int\limits_0^t |H_s(\omega)|\,d\,\mathrm{Var}\,(A)_s(\omega) < \infty, \\ +\infty, & \text{otherwise.} \end{cases} \qquad [2.17]$$

LEMMA 2.2.– Let A be a process with finite variation in the wide sense and H be a measurable stochastic process. Then, for every t, the mapping $\omega \rightsquigarrow Y_t(\omega)$, given by [2.17], is a random variable (with values in $(-\infty, +\infty]$). If, moreover, $A \in \mathcal{V}$ and H is progressively measurable, then $Y = (Y_t)_{t \in \mathbb{R}_+}$ is an adapted stochastic process with values in $(-\infty, +\infty]$.

PROOF.– It is enough to prove only the second assertion, because the first assertion follows from it if we put $\mathscr{F}_t \equiv \mathscr{F}$. Thus, let $A \in \mathcal{V}$. Fix $t \in \mathbb{R}_+$ and introduce the class $\mathscr{H}$ of bounded functions $H(\omega, s)$ on $\Omega \times [0, t]$, which are measurable with respect to the σ-algebra $\mathscr{F}_t \otimes \mathscr{B}([0, t])$ and such that the integral (taking finite values for every ω)

$$\int\limits_0^t H(\omega, s) \, dA_s(\omega),$$

is $\mathscr{F}_t$-measurable. Recall that this integral (and the first integral in [2.17] assuming the second integral is finite) is defined for every ω as the difference of the Lebesgue–Stieltjes integrals of the function $s \rightsquigarrow H_s(\omega)$ with respect to the Lebesgue–Stieltjes measures corresponding to the increasing functions $s \rightsquigarrow B_s(\omega)$ and $s \rightsquigarrow C_s(\omega)$, where B and C are from proposition 2.3. It is easy to see that $\mathscr{H}$ satisfies the assumptions of theorem A.3 on monotone classes. However, obviously, $\mathscr{H}$ contains the family $\mathscr{C}$ of functions H of the form $H = \mathbb{1}_{D \times \{0\}}$ or $H = \mathbb{1}_{D \times]u, v]}$, where $D \in \mathscr{F}_t$, $0 \leqslant u < v \leqslant t$. Since the family $\mathscr{C}$ generates the σ-algebra $\mathscr{F}_t \otimes \mathscr{B}([0, t])$, by the monotone class theorem, $\mathscr{H}$ consists of all bounded $\mathscr{F}_t \otimes \mathscr{B}([0, t])$-measurable functions.

Let a process H in the statement of the lemma be nonnegative and $A \in \mathcal{V}^+$. Apply the statement that was just proved to the functions $H \wedge n$ and pass to the limit as $n \to \infty$. The monotone convergence theorem shows that Y_t are $\mathscr{F}_t$-measurable (note that $Y_t(\omega) = \int_0^t H_s(\omega) \, dA_s(\omega)$ for all ω in this case).

In the general case, we now have that four integrals $\int_0^t H_s^+(\omega) \, dB_s(\omega)$, $\int_0^t H_s^-(\omega) \, dB_s(\omega)$, $\int_0^t H_s^+(\omega) \, dC_s(\omega)$, $\int_0^t H_s^-(\omega) \, dC_s(\omega)$ are $\mathscr{F}_t$-measurable. The claim follows easily.   □

THEOREM 2.11.– Let A be a process with finite variation in the wide sense and H be a measurable stochastic process. Assume that

$$P\left(\omega : \int\limits_0^t |H_s(\omega)| \, d\operatorname{Var}(A)_s(\omega) < \infty\right) = 1 \qquad [2.18]$$

for every $t \in \mathbb{R}_+$. Define $Y = (Y_t)_{t \in \mathbb{R}_+}$ as in [2.17]. There exists a process with finite variation in the wide sense and indistinguishable with Y. Any such process is denoted by $H \cdot A = (H \cdot A_t)_{t \in \mathbb{R}_+}$. Processes

$$\mathrm{Var}\,(H \cdot A) \quad \text{and} \quad |H| \cdot \mathrm{Var}\,(A) \quad \text{are indistinguishable.} \qquad [2.19]$$

Processes

$$\Delta(H \cdot A) \quad \text{and} \quad H\Delta A \quad \text{are indistinguishable.} \qquad [2.20]$$

If $A \in \mathscr{V}$ and H is progressively measurable, then $H \cdot A \in \mathscr{V}$. If A and H are predictable, then $H \cdot A$ is predictable.

REMARK 2.7.– The set in [2.18] is in $\mathscr{F}$ by lemma 2.2.

PROOF.– Put

$$D := \bigcap_{n=1}^{\infty} \left\{ \omega : \int_0^n |H_s(\omega)|\, d\,\mathrm{Var}\,(A)_s(\omega) < \infty \right\}$$

and $\widetilde{H} := H \mathbb{1}_{D \times \mathbb{R}_+}$. Define a process $\widetilde{Y}$ as in [2.17] with $\widetilde{H}$ instead of H. Then $\widetilde{Y} = Y \mathbb{1}_{D \times \mathbb{R}_+}$. By the assumptions, $\mathsf{P}(D) = 1$, hence, $\widetilde{Y}$ is indistinguishable with Y. Moreover,

$$\int_0^t |\widetilde{H}_s(\omega)|\, d\,\mathrm{Var}\,(A)_s(\omega) < \infty$$

and

$$\widetilde{Y}(\omega) = \int_0^t \widetilde{H}_s(\omega)\, dA_s(\omega)$$

for all ω and all t. Consequently, we can take $\widetilde{Y}$ as $H \cdot A$. Indeed, $\widetilde{Y}$ is a stochastic process by lemma 2.2. The required properties of trajectories of $\widetilde{Y}$ and relation [2.19] follow from properties of the Lebesgue–Stieltjes integral; see Appendix A.4. Further,

$$\Delta\widetilde{Y} = \widetilde{H}\Delta A, \qquad [2.21]$$

thus, [2.20] holds for this and, therefore, for every version of the process $H \cdot A$.

Note that the process $\mathbb{1}_{D \times \mathbb{R}_+}$ is predictable. Indeed, it has continuous trajectories and it is adapted due to the completeness of the filtration. Therefore, the process $\widetilde{H}$ is progressively measurable (respectively predictable), when H is progressively measurable (respectively predictable). Hence, when $A \in \mathcal{V}$ and H is progressively measurable, $\widetilde{Y}$ is adapted by lemma 2.2, where any version of $H \cdot A$ is adapted due to the completeness of the stochastic basis, i.e. $H \cdot A \in \mathcal{V}$.

Finally, let A and H be predictable. Then $\widetilde{Y}$ is predictable in view of [2.21] and proposition 1.8. An arbitrary version of $H \cdot A$ is predictable by corollary 1.2. $\square$

Thus, under the assumptions of theorem 2.11, given a pair (H, A) of processes, we have constructed an integral process $H \cdot A$. It is easy to see that this operation deals, essentially, with equivalence classes of indistinguishable stochastic processes: if H and H' are indistinguishable measurable processes, A and A' are indistinguishable processes with finite variation in the wide sense, H and A satisfy the assumptions of theorem 2.11, then H' and A' do the same and any version of the process $H' \cdot A'$ is a version of the process $H \cdot A$ and vice versa.

In the next chapter, we will introduce a similar notation for *stochastic* integrals. When it may cause confusion, we will use the notation $H \overset{s}{\cdot} A$ for the Lebesgue–Stieltjes integrals.

Let A be an increasing process in the wide sense and H be a nonnegative measurable process. Then the version $\widetilde{Y}$ of the process $H \cdot A$, constructed in the proof of theorem 2.11, is an increasing process in the wide sense. Therefore, for every version of $H \cdot A$, there exists, a.s., a limit $\lim_{t \uparrow \infty} H \cdot A_t = H \cdot A_\infty$, and, for almost all ω,

$$H \cdot A_\infty(\omega) = \int_0^\infty H_s(\omega) \, dA_s(\omega), \qquad [2.22]$$

where the integral on the right is understood as the Lebesgue–Stieltjes integral.

DEFINITION 2.10.– A stochastic process H is called *locally bounded* if there exists a localizing sequence $\{T_n\}$ of stopping times such that the process $H \mathbb{1}_{]0,T_n]}$ is bounded (uniformly in t and ω) for every n.

EXERCISE 2.23.– Let a process H be measurable and locally bounded, and let A be a process with finite variation in the wide sense. Show that [2.18] holds for every $t \in \mathbb{R}_+$.

In the following, we are especially interested in predictable integrands. It is convenient to introduce a corresponding notation.

DEFINITION 2.11.– Let $A \in \mathcal{V}$. Then $L_{\mathrm{var}}(A)$ denotes the class of all *predictable* processes H satisfying [2.18] for every $t \in \mathbb{R}_+$.

EXERCISE 2.24.– Let H be a measurable process indistinguishable with zero process. Show that $H \in L_{\mathrm{var}}(A)$ for every $A \in \mathcal{V}$.

HINT.– Use proposition 1.12.

For ease of reference, we state the following theorem. All equalities are understood up to indistinguishability. Statements that have not been proved before are left to the reader as an exercise.

THEOREM 2.12.– Let $A \in \mathcal{V}$ and $H \in L_{\mathrm{var}}(A)$. Then $H \cdot A \in \mathcal{V}$ and the following statements hold:

1) if $B \in \mathcal{V}$, $H \in L_{\mathrm{var}}(B)$ and $\alpha, \beta \in \mathbb{R}$, then $H \in L_{\mathrm{var}}(\alpha A + \beta B)$ and

$$H \cdot (\alpha A + \beta B) = \alpha(H \cdot A) + \beta(H \cdot B);$$

2) if $K \in L_{\mathrm{var}}(A)$ and $\alpha, \beta \in \mathbb{R}$, then $\alpha H + \beta K \in L_{\mathrm{var}}(A)$ and

$$(\alpha H + \beta K) \cdot A = \alpha(H \cdot A) + \beta(K \cdot A);$$

3) if T is a stopping time, then $H \in L_{\mathrm{var}}(A^T)$, $H \mathbb{1}_{\rrbracket 0,T \rrbracket} \in L_{\mathrm{var}}(A)$ and

$$(H \cdot A)^T = H \cdot A^T = \left(H \mathbb{1}_{\rrbracket 0,T \rrbracket}\right) \cdot A;$$

4) $K \in L_{\mathrm{var}}(H \cdot A) \iff KH \in L_{\mathrm{var}}(A)$ and then $K \cdot (H \cdot A) = (KH) \cdot A$;

5) $\Delta(H \cdot A) = H \Delta A$;

6) if A is predictable, then $H \cdot A$ is predictable;

7) if $A \in \mathcal{V}^+$, $H \geqslant 0$, then there is a version $H \cdot A$ which lies in $\mathcal{V}^+$.

EXERCISE 2.25.– Prove statements (1)–(4) of the theorem.

2.4. Integrable increasing processes and processes with integrable variation. Doléans measure

We will assume that a stochastic basis $\mathbb{B} = (\Omega, \mathscr{F}, \mathbb{F} = (\mathscr{F}_t)_{t \in \mathbb{R}_+}, \mathsf{P})$ satisfying the usual conditions is given.

DEFINITION 2.12.– An increasing stochastic process $A = (A_t)_{t \in \mathbb{R}_+} \in \mathcal{V}^+$ is called an *integrable increasing process* if $\mathsf{E} A_\infty < \infty$. A process with finite variation $A = (A_t)_{t \in \mathbb{R}_+} \in \mathcal{V}$ is called a *process with integrable variation* if $\mathsf{E} \mathrm{Var}\,(A)_\infty < \infty$.

The class of all integrable increasing processes is denoted by $\mathscr{A}^+$, and the class of all processes with integrable variation is denoted by $\mathscr{A}$.

Similarly to the previous section, an integrable increasing process in the wide sense (respectively a process with integrable variation in the wide sense) is understood as an integrable increasing process (respectively a process with integrable variation) relative to the filtration $\mathscr{F}_t \equiv \mathscr{F}$.

It is clear that $\mathscr{A}^+$ and $\mathscr{A}$ are stable under stopping and $\mathscr{A} = \mathscr{A}^+ - \mathscr{A}^+$. If $A \in \mathscr{A}$ (or A is a process with integrable variation in the wide sense), then, for ω such that $\mathrm{Var}\,(A)_\infty(\omega) < \infty$, there exists a finite limit $\lim_{t \uparrow \infty} A_t(\omega)$, so that a random variable A_∞ is defined and finite a.s.

It is possible to associate a finite measure on the σ-algebra of measurable sets with an integrable increasing process in the wide sense. This measure is called the Doléans measure. Correspondingly, given a process with integrable variation in the wide sense, we associate a bounded signed measure on the σ-algebra of measurable sets with it, which is also called the Doléans measure. Moreover, optional (i.e. adapted) integrable increasing processes are indistinguishable if the corresponding Doléans measures coincide on the optional σ-algebra, while predictable integrable increasing processes are indistinguishable if the corresponding Doléans measures coincide on the predictable σ-algebra. A more complicated important question is whether a measure on the optional (respectively predictable) σ-algebra is the restriction onto this σ-algebra of the Doléans measure of some adapted (respectively predictable) integrable increasing process. This section is devoted to the study of these issues.

DEFINITION 2.13.– Let A be a process with integrable variation in the wide sense. A signed measure μ_A on the space $(\Omega \times \mathbb{R}_+, \mathscr{F} \otimes \mathscr{B}(\mathbb{R}_+))$ defined by

$$\mu_A(B) = \mathsf{E}(\mathbb{1}_B \cdot A_\infty),$$

is called the *Doléans measure* corresponding to A.

PROPOSITION 2.4.– Definition 2.13 is correct. The signed measure μ_A takes finite values and vanishes on evanescent sets. If A is increasing in the wide sense process, then μ_A is nonnegative.

PROOF.– Obviously, $\mathbb{1}_B \cdot A$ is a process with integrable variation in the wide sense for every $B \in \mathscr{F} \otimes \mathscr{B}(\mathbb{R}_+)$. Hence, $\mu_A(B)$ is defined and finite; moreover, it is equal to 0 on evanescent sets and takes nonnegative values if A is an increasing process in the wide sense. So it is enough to check that μ_A is countably additive and only for an increasing A. Let B be the union of a countable number of pairwise disjoint sets $B_n \in \mathscr{F} \otimes \mathscr{B}(\mathbb{R}_+)$. Put

$$\xi_n(\omega) = \mathbb{1}_{B_n} \cdot A_\infty(\omega), \qquad \xi(\omega) = \mathbb{1}_B \cdot A_\infty(\omega).$$

In view of [2.22] and since the Lebesgue–Stieltjes measure corresponding to the function $t \rightsquigarrow A_t(\omega)$ is countably additive, we have $\xi = \sum_n \xi_n$ a.s. By the monotone (or dominated) convergence theorem,

$$\mu_A(B) = \mathsf{E}\xi = \sum_{n=1}^{\infty} \mathsf{E}\xi_n = \sum_{n=1}^{\infty} \mu_A(B_n). \qquad \square$$

PROPOSITION 2.5.– Let A be a process with integrable variation in the wide sense and H be a bounded measurable process. Then the following integrals are defined, finite and

$$\int_{\Omega \times \mathbb{R}_+} H \, d\mu_A = \mathsf{E}(H \cdot A_\infty).$$

EXERCISE 2.26.– Prove proposition 2.5.

Using the Doléans measure, it is easy to characterize those processes in $\mathscr{A}$ that are martingales. Before we state the result, let us make a trivial remark that will be used many times below without explanation. For $A \in \mathscr{A}$, we have $A_\infty^* \leqslant \mathrm{Var}\,(A)_\infty \in L^1$, Hence, every local martingale in $\mathscr{A}$ is a uniformly integrable martingale and, moreover, belongs to $\mathscr{H}^1$.

THEOREM 2.13.– A process $A \in \mathscr{A}$ is a martingale if and only if $\mu_A(B) = 0$ for every $B \in \mathscr{P}$.

PROOF.– To prove the sufficiency, it is enough to note that sets $B = C \times]s, t]$, $C \in \mathscr{F}_s$, $0 \leqslant s < t$, are predictable, so $\mu_A(B) = 0$ implies

$$\mu_A(B) = \mathsf{E}\mathbb{1}_C(A_t - A_s) = 0,$$

which signifies that $\mathsf{E}(A_t | \mathscr{F}_s) = A_s$ a.s.

Let $A \in \mathscr{A} \cap \overline{\mathscr{M}}$. In view of the remark before the theorem, $\mathsf{E}A_\infty = \mathsf{E}A_0 = 0$, i.e. $\mu_A(\Omega \times \mathbb{R}_+) = 0$. It follows from the above formula that μ_A vanishes on sets of the form $B = C \times]s, t]$, $C \in \mathscr{F}_s$, $0 \leqslant s < t$. It is also obvious that $\mu_A(B) = 0$ if B has the form $B = C \times \{0\}$, $C \in \mathscr{F}_0$. Sets B of two indicated forms constitute a π-system generating the σ-algebra $\mathscr{P}$ (theorem 1.10). It remains to note that the collection of sets on which a signed measure vanishes, satisfies conditions 2) and 3) in the definition of a λ-system, and then to use theorem A.2 on π-λ-systems. $\quad \square$

THEOREM 2.14.– Let A and A' be the processes with integrable variation. If $\mu_A(B) = \mu_{A'}(B)$ for every $B \in \mathscr{O}$, then A and A' are indistinguishable.

THEOREM 2.15.– Let A and A' be the predictable processes with integrable variation. If $\mu_A(B) = \mu_{A'}(B)$ for every $B \in \mathscr{P}$, then A and A' are indistinguishable.

The following important result is an immediate consequence of theorems 2.15 and 2.13.

COROLLARY 2.10.– A predictable martingale in $\mathscr{A}$ is indistinguishable with the zero process.

The proof of theorems 2.14 and 2.15 is based on the following lemma, whose proof is given after we prove the theorems. For brevity, hereinafter we say that $M = (M_t)$ is a martingale *associated* with a bounded random variable ξ if M is a bounded martingale such that $M_t = E(\xi | \mathscr{F}_t)$ a.s. for every $t \in \mathbb{R}_+$.

LEMMA 2.3.– Let $A \in \mathscr{A}$, ξ a bounded random variable, M a bounded martingale associated with ξ, T a stopping time. Then

$$E\xi A_T = E(M \cdot A_T). \qquad [2.23]$$

If, moreover, A is predictable, then

$$E\xi A_T = E(M_- \cdot A_T). \qquad [2.24]$$

PROOF OF THEOREM 2.14.– Take any $t \in \mathbb{R}_+$ and an arbitrary set $B \in \mathscr{F}$. Let M be a martingale associated with $\mathbb{1}_B$. It follows from [2.23] and proposition 2.5 that

$$E\mathbb{1}_B A_t = E(M \cdot A_t) = E(M\mathbb{1}_{[0.t]} \cdot A_\infty) = \int\limits_{\Omega \times \mathbb{R}_+} M\mathbb{1}_{[0.t]} \, d\mu_A.$$

Similarly,

$$E\mathbb{1}_B A_t' = \int\limits_{\Omega \times \mathbb{R}_+} M\mathbb{1}_{[0.t]} \, d\mu_{A'}.$$

Since signed measures μ_A and $\mu_{A'}$ coincide on the optional σ-algebra, integrals of arbitrary optional functions with respect to these measures also coincide. Hence, $E\mathbb{1}_B A_t = E\mathbb{1}_B A_t'$. Since $B \in \mathscr{F}$ is arbitrary, we have $A_t = A_t'$ a.s., i.e. A and A' are modifications of each other. Indistinguishability of A and A' follows from proposition 1.1. □

PROOF OF THEOREM 2.15.– This is similar to the proof of theorem 2.14. We should only use [2.24] instead of [2.23], which allows us to replace the optional process M by the predictable process M_- in the displayed formulas. □

PROOF OF LEMMA 2.3.– Let us first prove [2.23] assuming that $A \in \mathscr{A}^+$ and $T \equiv t \in \mathbb{R}_+$. Since M is bounded and right-continuous, P-a.s.,

$$M \cdot A_t = \lim_{n \to \infty} \sum_{k=1}^{n} M_{\frac{k}{n}t} \left(A_{\frac{k}{n}t} - A_{\frac{k-1}{n}t} \right)$$

by the Lebesgue dominated convergence theorem. Using this theorem again, we get

$$\mathsf{E}(M \cdot A_t) = \lim_{n \to \infty} \sum_{k=1}^{n} \mathsf{E} \left\{ M_{\frac{k}{n}t} \left(A_{\frac{k}{n}t} - A_{\frac{k-1}{n}t} \right) \right\}$$

$$= \lim_{n \to \infty} \sum_{k=1}^{n} \mathsf{E} \left\{ \mathsf{E}(\xi | \mathscr{F}_{\frac{k}{n}t}) \left(A_{\frac{k}{n}t} - A_{\frac{k-1}{n}t} \right) \right\}$$

$$= \lim_{n \to \infty} \sum_{k=1}^{n} \mathsf{E} \left(\xi \left(A_{\frac{k}{n}t} - A_{\frac{k-1}{n}t} \right) \right) = \mathsf{E}\xi A_t.$$

Now we prove [2.24] under the same assumptions as before and, additionally, for a predictable A. In view of [2.23], it is sufficient to check that $\mathsf{E}(\Delta M) \cdot A_t = 0$. By theorem 1.18, there is a sequence $\{T_n\}$ of predictable stopping times, exhausting jumps of A, i.e. such that $[\![T_n]\!] \cap [\![T_m]\!] = \varnothing$ for $m \neq n$ and $\{\Delta A \neq 0\} = \cup_n [\![T_n]\!]$. Since, for a fixed ω, the function $t \rightsquigarrow \Delta M_t(\omega)$ takes at most a countable number of values, we have, P-a.s.,

$$(\Delta M) \cdot A_t = \sum_{n=1}^{\infty} \Delta M_{T_n} \Delta A_{T_n} \mathbb{1}_{\{T_n \leqslant t\}}.$$

Applying the dominated convergence theorem, proposition 1.11 and theorem 2.4, we get

$$\mathsf{E}(\Delta M) \cdot A_t = \sum_{n=1}^{\infty} \mathsf{E} \left\{ \Delta M_{T_n} \Delta A_{T_n} \mathbb{1}_{\{T_n \leqslant t\}} \right\}$$

$$= \sum_{n=1}^{\infty} \mathsf{E} \left\{ \Delta A_{T_n} \mathbb{1}_{\{T_n \leqslant t\}} \mathsf{E} \left(\Delta M_{T_n} \mathbb{1}_{\{T_n < \infty\}} | \mathscr{F}_{T_n -} \right) \right\} = 0.$$

Thus, [2.23] and [2.24] are proved in the case where T is identically equal to a finite t. Passing to the limit as $t \to \infty$ and using the dominated convergence theorem, we get the result for $T \equiv \infty$. If T is an arbitrary stopping time, we apply the statement we have just proved (with $T \equiv \infty$) to the stopped process A^T instead of A. Using theorem 2.12 (3), we prove [2.23] and [2.24] for an arbitrary T. Finally, the general case with $A \in \mathscr{A}$ reduces to the considered case with $A \in \mathscr{A}^+$ due to proposition 2.3.

$\square$

The assumptions that A is optional or predictable were used in the proofs of theorems 2.14 and 2.15 only through relations [2.23] and [2.24]. Thus, it is natural to ask whether [2.23] and [2.24] are valid for a broader class of processes with integrable variation in the wide sense than optional or predictable processes, respectively. The answer to this question is negative, and it is easy to make sure of this in the optional case, while the predictable case is more difficult. Let us first consider the discrete-time case. Then [2.23] and [2.24] are interpreted (for $T \equiv n$) as

$$\mathsf{E}\xi A_n = \sum_{k=1}^{n} \mathsf{E}(\xi|\mathscr{F}_k)(A_k - A_{k-1}) \tag{2.25}$$

and

$$\mathsf{E}\xi A_n = \sum_{k=1}^{n} \mathsf{E}(\xi|\mathscr{F}_{k-1})(A_k - A_{k-1}) \tag{2.26}$$

respectively, where ξ is an arbitrary bounded random variable. Now it follows from [2.25] that $\mathsf{E}\xi A_n = 0$ for every bounded random variable ξ satisfying $\mathsf{E}(\xi|\mathscr{F}_n) = 0$. Due to exercise 2.27 given below, this means that A_n a.s. coincides with an $\mathscr{F}_n$-measurable random variable (and is $\mathscr{F}_n$-measurable itself if the stochastic basis is complete). Similarly, it follows from [2.26] that $\mathsf{E}\xi A_n = 0$ for every bounded random variable ξ satisfying $\mathsf{E}(\xi|\mathscr{F}_{n-1}) = 0$, i.e. A_n a.s. coincides with an $\mathscr{F}_{n-1}$-measurable random variable.

EXERCISE 2.27.– Let $(\Omega, \mathscr{F}, \mathsf{P})$ be a probability space, $\mathscr{G}$ a sub-σ-algebra of the σ-algebra $\mathscr{F}$ and η be an integrable random variable. Assume that $\mathsf{E}\eta\xi = 0$ for every bounded random variable ξ satisfying $\mathsf{E}(\xi|\mathscr{G}) = 0$. Show that $\eta = \mathsf{E}(\eta|\mathscr{G})$ P-a.s.

HINT.– Apply the Hahn–Banach theorem.

In the continuous-time case, the statement that [2.23] implies that A is optional, is proved similarly. We leave the details to the reader as an exercise.

EXERCISE 2.28.– Let A be a process with integrable variation in the wide sense. Assume that, for every bounded random variable ξ and for every $t \in \mathbb{R}_+$, [2.23] holds for $T \equiv t$, where M is a martingale associated with ξ. Prove that $A \in \mathscr{A}$.

With regard to the predictable case, we can prove similarly that if [2.24] holds for a predictable stopping time T, then A_T is $\mathscr{F}_{T-}$-measurable (see details of the proof later on). However, this is not enough to prove the predictability of A; see theorem 1.19. For the time being, let us introduce the following definition.

DEFINITION 2.14.– A process A with integrable variation in the wide sense is called *natural* if, for every bounded random variable ξ and for every $t \in \mathbb{R}_+$,

$$\mathsf{E}\xi A_t = \mathsf{E}(M_- \cdot A_t), \qquad\qquad [2.27]$$

where M is a martingale associated with ξ.

The second part of lemma 2.3 says that a predictable process with integrable variation is natural. As already been mentioned, the converse statement will be proved later. Historically, the notion of a natural increasing process appeared before the notion of the predictable σ-algebra.

EXERCISE 2.29.– Show that a natural increasing process is adapted, i.e. belongs to $\mathscr{A}^+$.

Recall that the optional and predictable projections are considered in section 1.6.

PROPOSITION 2.6.– Let $A \in \mathscr{A}^+$. Then, for every bounded measurable process X, the optional projection OX is a version of the "conditional expectation" $\mathsf{E}_{\mu_A}(X|\mathscr{O})$. In particular,

$$\int_{\Omega \times \mathbb{R}_+} (OX)\, d\mu_A = \int_{\Omega \times \mathbb{R}_+} X\, d\mu_A. \qquad\qquad [2.28]$$

If, additionally, A is a natural process, then, for every bounded measurable process X, the predictable projection ΠX is a version of the "conditional expectation" $\mathsf{E}_{\mu_A}(X|\mathscr{P})$. In particular,

$$\int_{\Omega \times \mathbb{R}_+} (\Pi X)\, d\mu_A = \int_{\Omega \times \mathbb{R}_+} X\, d\mu_A. \qquad\qquad [2.29]$$

REMARK 2.8.– "Conditional expectations" $\mathsf{E}_{\mu_A}(X|\mathscr{O})$ and $\mathsf{E}_{\mu_A}(X|\mathscr{P})$ (we use the quotes because μ_A is a finite measure but not, in general, a probability measure) are defined up to μ_A-null sets. These include evanescent sets, however, the class of μ_A-null sets is much wider. For example, let $A = \mathbb{1}_{[\![T,\infty[\![}$, where T is a stopping time. Then $\mu_A(B) = 0$ means that the set $B \cap [\![T]\!]$ is evanescent. That is why the projections cannot be defined as the corresponding conditional expectations.

PROOF.– We prove only the second statement; the optional case is handled similarly. We must prove that, for every bounded measurable process X and for every $B \in \mathscr{P}$,

$$\int_{\Omega \times \mathbb{R}_+} (\Pi X)\mathbb{1}_B\, d\mu_A = \int_{\Omega \times \mathbb{R}_+} X\mathbb{1}_B\, d\mu_A.$$

Since $\mathbb{1}_B$ can be put under the sign of predictable projection because of the third equality in [1.8], it is enough to consider the case where $B = \Omega$, i.e. to check that [2.29] holds for every bounded measurable process X.

Now we can see that [2.27] in the definition of a natural process means that [2.29] is valid for processes X of the form $X = \xi \mathbb{1}_{[0,t]}$, where ξ is a bounded random variable; see the expression for ΠX for such X in the proof of theorem 1.21.

Let $\mathcal{H}$ be the set of bounded measurable processes X for which [2.29] is true. Clearly, $\mathcal{H}$ is a linear space and contains constant functions. If $0 \leqslant X^1 \leqslant \ldots \leqslant X^n \leqslant \ldots \leqslant C$, $X^n \in \mathcal{H}$ and $X = \lim_n X^n$, then, as we know from the proof of theorem 1.21, $\Pi X = \lim_n \Pi X^n$ up to an evanescent set. Therefore, by the Lebesgue dominated convergence theorem, we can pass to the limit under the integral signs in [2.29] for X^n as $n \to \infty$. Hence, $X \in \mathcal{H}$. Now the claim follows from theorem A.3 on monotone classes. $\qquad\square$

Now we can prove the desired result.

THEOREM 2.16.– A natural process with integrable variation in the wide sense is predictable.

PROOF.– The proof is based on theorem 1.19. Let A be a natural process with integrable variation in the wide sense.

Let T be a predictable stopping time, and let ξ be a bounded random variable such that $\mathsf{E}(\xi | \mathscr{F}_{T-}) = 0$. Consider a bounded martingale M associated with ξ. We assert that $M_- \mathbb{1}_{[0,T]} = 0$.

Indeed, let $B \in \mathscr{F}_t$ and $t \in \mathbb{R}_+$. Then $B \cap \{t < T\} \in \mathscr{F}_{T-} \cap \mathscr{F}_t$. Hence,

$$\int_B M_t \mathbb{1}_{\{t<T\}} \, d\mathsf{P} = \int \mathsf{E}(\xi | \mathscr{F}_t) \mathbb{1}_B \mathbb{1}_{\{t<T\}} \, d\mathsf{P} = \int \xi \mathbb{1}_B \mathbb{1}_{\{t<T\}} \, d\mathsf{P}$$

$$= \int \mathsf{E}(\xi | \mathscr{F}_{T-}) \mathbb{1}_B \mathbb{1}_{\{t<T\}} \, d\mathsf{P} = 0.$$

Therefore, $M_t \mathbb{1}_{\{t<T\}} = 0$ a.s. and, for ω from a set of measure one, $M_t(\omega) \mathbb{1}_{\{t<T(\omega)\}} = 0$ simultaneously for all nonnegative rational t. Since trajectories $M_{\cdot}(\omega)$ are right-continuous, we conclude that the process $M \mathbb{1}_{[0,T[}$ is indistinguishable with the zero process. Hence, $M_- \mathbb{1}_{[0,T]} \mathbb{1}_{\{T>0\}}$ is indistinguishable with the zero process. It remains to note that $M_0 = \mathsf{E}(\xi | \mathscr{F}_0) = \mathsf{E}(\mathsf{E}(\xi | \mathscr{F}_{T-}) | \mathscr{F}_0) = 0$ a.s.

Put $X := \xi \mathbb{1}_{[0,T]}$. Then

$$\Pi X = \Pi(\xi \mathbb{1}_{[0,\infty[}) \mathbb{1}_{[0,T]} = M_- \mathbb{1}_{[0,T]} = 0.$$

By [2.29], we get

$$0 = \int_{\Omega \times \mathbb{R}_+} X \, d\mu_A = \mathsf{E}\xi A_T.$$

According to exercise 2.27, the random variable A_T is $\mathscr{F}_{T-}$-measurable.

Let S be a totally inaccessible stopping time and $B \in \mathscr{F}_S$. Put $X = \mathbb{1}_{[\![S_B]\!]}$. It follows from the definitions of predictable projection and totally inaccessible stopping time that $\Pi X = 0$. Using [2.29], we get

$$0 = \int_{\Omega \times \mathbb{R}_+} X \, d\mu_A = \mathsf{E}\Delta A_S \mathbb{1}_{\{S<\infty\}} \mathbb{1}_B.$$

Since the random variable $\Delta A_S \mathbb{1}_{\{S<\infty\}}$ is $\mathscr{F}_S$-measurable, we obtain

$$\mathsf{P}(S < \infty, \, \Delta A_S \neq 0) = 0. \qquad \Box$$

THEOREM 2.17.– Let μ be a finite measure on the σ-algebra $\mathscr{P}$ such that $\mu([\![0]\!]) = 0$ and $\mu(B) = 0$ if B is evanescent. Then there exists a predictable increasing process $A \in \mathscr{A}^+$ such that its Doléans measure coincides with μ on $\mathscr{P}$.

PROOF.– Take $t \in \mathbb{R}_+$ and $B \in \mathscr{F}_t$. Let $M = (M_t)_{t\in\mathbb{R}_+}$ be a bounded martingale associated with $\mathbb{1}_B$. Define

$$\nu_t(B) = \int_{\Omega \times \mathbb{R}_+} M_- \mathbb{1}_{[\![0,t]\!]} \, d\mu.$$

Since μ vanishes on evanescent sets, the expression on the right does not depend on the choice of a version of M. It is obvious that ν_t is finitely additive on $\mathscr{F}_t$. To prove that it is countably additive, let us check that it is continuous at $\varnothing$. Let $B_n \in \mathscr{F}_t$, $B_1 \supseteq B_2 \supseteq \cdots \supseteq B_n \supseteq \ldots$, $\cap_n B_n = \varnothing$, and let M^n be martingales associated with B_n; without loss of generality, we may assume that $1 \geqslant M_s^1(\omega) \geqslant M_s^2(\omega) \geqslant \ldots \geqslant M_s^n(\omega) \geqslant \ldots \geqslant 0$ for all ω and s. By Doob's inequality (corollary 2.6 and remark 2.5)

$$\mathsf{E}\big((M^n)_\infty^*\big)^2 \leqslant 4\mathsf{E}|M_\infty^n|^2 = 4\mathsf{P}(B_n) \to 0, \quad n \to \infty.$$

Hence, $(M^n)_\infty^* \xrightarrow{P} 0$ and, since the sequence M^n is monotonic in n, $(M^n)_\infty^* \to 0$ a.s. Thus, we have the monotone convergence of the sequence $M_-^n \mathbb{1}_{[\![0,t]\!]}$ to a process indistinguishable with zero process. Therefore, $\nu_t(B_n) \to 0$ by the monotone convergence theorem, and we conclude that ν_t is a finite measure on $\mathscr{F}_t$.

Obviously, ν_t is absolutely continuous with respect to the restriction P_t of P onto $\mathscr{F}_t$. Put

$$A'_t := \frac{d\nu_t}{d\mathsf{P}_t}.$$

By the definition, A'_t is an $\mathscr{F}_t$-measurable nonnegative random variable and $\mathsf{E}A'_t = \mu(\llbracket 0, t \rrbracket)$. In particular, $A'_0 = 0$ a.s. Moreover, for every bounded $\mathscr{F}_t$-measurable random variable ξ,

$$\mathsf{E}\xi A'_t = \int_\Omega \xi \, d\nu_t = \int_{\Omega \times \mathbb{R}_+} N_- \mathbb{1}_{\llbracket 0, t \rrbracket} \, d\mu, \qquad\qquad [2.30]$$

where N is a martingale associated with ξ. Indeed, [2.30] holds for indicator functions and, hence, their linear combinations, from the definition of A'_t. Then it is proved for nonnegative bounded ξ by passing to a limit in monotone sequences of linear combinations of indicators similarly to the proof of continuity of ν_t at $\varnothing$. Finally, use the decomposition $\xi = \xi^+ - \xi^-$ in the general case.

Let $0 \leqslant s < t$, $B \in \mathscr{F}_t$, and let M be a martingale associated with $\mathbb{1}_B$ as above. Then

$$\mathsf{E}\mathbb{1}_B A'_t = \int_{\Omega \times \mathbb{R}_+} M_- \mathbb{1}_{\llbracket 0, t \rrbracket} \, d\mu$$

by the definitions and

$$\mathsf{E}\mathbb{1}_B A'_s = \mathsf{E}\big[\mathsf{E}(\mathbb{1}_B | \mathscr{F}_s) A'_s\big] = \int_{\Omega \times \mathbb{R}_+} M_- \mathbb{1}_{\llbracket 0, s \rrbracket} \, d\mu$$

in view of [2.30]. Here we use that a martingale N associated with $\xi = \mathsf{E}(\mathbb{1}_B | \mathscr{F}_s)$ is indistinguishable with M on $\llbracket 0, s \rrbracket$. Therefore, $\mathsf{E}\mathbb{1}_B(A'_t - A'_s) \geqslant 0$ for every $B \in \mathscr{F}_t$, hence

$$A'_t \geqslant A'_s \quad \text{a.s.} \qquad\qquad [2.31]$$

Put, for $t \in \mathbb{R}_+$,

$$A''_t = \inf_{r \in \mathbb{Q}:\, r > t} A'_r.$$

By the construction, for every ω, the function $t \rightsquigarrow A_t''(\omega)$ takes nonnegative values, is nondecreasing and right-continuous. Next, for every $t \in \mathbb{R}_+$, we have $A_t'' \geqslant A_t'$ a.s. in view of [2.31]. However,

$$\mathsf{E}A_t'' \leqslant \inf_{r \in \mathbb{Q}:\, r>t} \mathsf{E}A_r' = \inf_{r \in \mathbb{Q}:\, r>t} \mu([\![0,r]\!]) = \mu([\![0,t]\!]) = \mathsf{E}A_t'.$$

Hence, $A_t'' = A_t'$ a.s. In particular, A_t'' are $\mathscr{F}_t$-measurable and $A_0'' = 0$ a.s.

Now define

$$A_t := A_t'' \mathbb{1}_{\{A_0''=0\}}, \quad t \in \mathbb{R}_+.$$

It follows from the above listed properties of the process A'' that $A = (A_t)_{t \in \mathbb{R}_+} \in \mathscr{V}^+$. Moreover, $A \in \mathscr{A}^+$ because, for every $t \in \mathbb{R}_+$,

$$\mathsf{E}A_t = \mathsf{E}A_t'' = \mathsf{E}A_t' = \mu([\![0,t]\!]) \leqslant \mu([\![0,\infty[\![).$$

Let $0 \leqslant s < t$, $B \in \mathscr{F}_s$, and let M be a martingale associated with $\mathbb{1}_B$. Then M is indistinguishable with $\mathbb{1}_B$ on $[\![s,\infty[\![$ and, hence, M_- is indistinguishable with $\mathbb{1}_B$ on $]\!]s,\infty[\![$. Therefore,

$$\mu_A(B \times]s,t]) = \mathsf{E}\mathbb{1}_B(A_t - A_s) = \mathsf{E}\mathbb{1}_B(A_t' - A_s')$$

$$= \int_{\Omega \times \mathbb{R}_+} M_- \mathbb{1}_{]s,t]}\, d\mu = \mu(B \times]s,t]).$$

That μ_A and μ coincide on $\mathscr{P}$ now follows from theorem 1.10 and from theorem A.2 on π-λ-systems.

Due to theorem 2.16, it remains to prove that A is a natural process. Let $t \in \mathbb{R}_+$, ξ be a bounded random variable, and let M be a martingale associated with ξ. Then a martingale N associated with a random variable M_t is indistinguishable with M on $[\![0,t]\!]$. Therefore,

$$\mathsf{E}\xi A_t = \mathsf{E}M_t A_t = \mathsf{E}M_t A_t' = \int_{\Omega \times \mathbb{R}_+} M_- \mathbb{1}_{[\![0,t]\!]}\, d\mu$$

$$= \int_{\Omega \times \mathbb{R}_+} M_- \mathbb{1}_{[\![0,t]\!]}\, d\mu_A = \mathsf{E}(M_- \cdot A_t). \qquad \square$$

It follows from proposition 2.6 and theorem 2.16 that, for every bounded measurable process X and for every predictable $A \in \mathscr{A}$,

$$\int_{\Omega \times \mathbb{R}_+} (\Pi X)\, d\mu_A = \int_{\Omega \times \mathbb{R}_+} X\, d\mu_A.$$

It turns out that it is possible to "change the roles" of X and A.

PROPOSITION 2.7.– Let A be a process with integrable variation in the wide sense. There exists a unique (up to indistinguishability) predictable process $B \in \mathscr{A}$ such that

$$\int_{\Omega \times \mathbb{R}_+} X\, d\mu_A = \int_{\Omega \times \mathbb{R}_+} X\, d\mu_B \qquad\qquad [2.32]$$

for every bounded predictable process X. This process is called the *dual predictable projection* of A and denoted by A^π.

COROLLARY 2.11.– Let $A \in \mathscr{A}^+$. There exists a unique (up to indistinguishability) predictable increasing process $\tilde{A} \in \mathscr{A}^+$ such that $A - \tilde{A} \in \mathscr{M}$.

PROOF OF PROPOSITION 2.7.– Relation [2.32] means that the Doléans measures of the processes A and B coincide on the predictable σ-algebra $\mathscr{P}$. Hence, the uniqueness follows from theorem 2.15, and the existence, in the case where A is an integrable increasing process in the wide sense, follows from theorems 2.17 and 2.16. The existence in the general case follows from the decomposition in proposition 2.3. □

DEFINITION 2.15.– The dual predictable projection A^π of a process $A \in \mathscr{A}$ with integrable variation is also called the *compensator* of A and often denoted by $\tilde{A}$.

LEMMA 2.4.– Let $A \in \mathscr{A}^+$, and let $\tilde{A}$ be the compensator of A. Then, for every predictable stopping time T,

$$\mathsf{E}(\Delta A_T \mathbb{1}_{\{T<\infty\}} | \mathscr{F}_{T-}) = \Delta \tilde{A}_T \mathbb{1}_{\{T<\infty\}}.$$

In other words, the assertion of the lemma can be formulated as follows:

$$\Pi(\Delta A) = \Delta \tilde{A},$$

see theorem 1.22 and remark 1.8.

PROOF.– Since $A - \tilde{A}$ is a uniformly integrable martingale, see corollary 2.11, the claim follows from theorem 2.4. □

THEOREM 2.18.– Let $A \in \mathscr{A}^+$. The compensator $\widetilde{A}$ of A is a.s. continuous if and only if $\Delta A_T \mathbb{1}_{\{T<\infty\}} = 0$ a.s. for every predictable stopping time T.

PROOF.– The necessity is evident because of lemma 2.4. To prove the sufficiency, let us use theorem 1.18 to represent the set $\{\Delta \widetilde{A} > 0\}$ as the union of the graphs of predictable stopping times T_n. Then, on the one hand, $\Delta \widetilde{A}_{T_n} > 0$ on the set $\{T_n < \infty\}$. On the other hand, due to the assumption and by lemma 2.4, $\Delta \widetilde{A}_{T_n} \mathbb{1}_{\{T_n<\infty\}} = 0$ a.s. Therefore, $\mathsf{P}(T_n < \infty) = 0$ for every n, hence the set $\{\Delta \widetilde{A} > 0\}$ is evanescent. $\square$

EXERCISE 2.30.– Let $X = (X_t)_{t \in \mathbb{R}_+}$ be a Poisson process on a stochastic basis $(\Omega, \mathscr{F}, \mathbb{F}, \mathsf{P})$. Put $T_n := \inf\{t \colon X_t \geqslant n\}$. Show that T_n is a totally inaccessible stopping time for every n.

HINT.– Apply theorem 2.18 to the process X^{T_n}.

EXERCISE 2.31.– Let μ be a finite measure on the σ-algebra $\mathscr{F} \otimes \mathscr{B}(\mathbb{R}_+)$ of measurable sets such that $\mu(\llbracket 0 \rrbracket) = 0$ and $\mu(B) = 0$ if B is evanescent. Prove that there exists an integrable increasing process in the wide sense A such that $\mu_A = \mu$.

Now we formulate a result dealing with the optional case which is not used later. The proof is omitted. It is similar to the proof of theorem 2.17, with M_- replaced by M in the definition of ν_t. However, that μ_A and μ coincide on $\mathscr{O}$ is proved more difficult because there is no convenient characterization of the optional σ-algebra $\mathscr{O}$ similar to that of predictable σ-algebra from theorem 1.10.

THEOREM 2.19.– Let μ be a finite measure on the σ-algebra $\mathscr{O}$ such that $\mu(\llbracket 0 \rrbracket) = 0$ and $\mu(B) = 0$ if B is evanescent. Then there exists an increasing process $A \in \mathscr{A}^+$ such that its Doléans measure coincides with μ on $\mathscr{O}$.

EXERCISE 2.32.– Prove theorem 2.19.

HINT.– Follow the scheme suggested above. To prove that μ and μ_A coincide on $\mathscr{O}$, show with the use of theorem A.3 on monotone classes, that $\int (OX)\,d\mu = \int (OX)\,d\mu_A$ for every bounded measurable process X.

PROPOSITION 2.8.– Let A be a process with integrable variation in the wide sense. There exists a unique (up to indistinguishability) process $B \in \mathscr{A}$ such that

$$\int_{\Omega \times \mathbb{R}_+} X\,d\mu_A = \int_{\Omega \times \mathbb{R}_+} X\,d\mu_B. \qquad [2.33]$$

for every bounded optional process X. This process is called the *dual optional projection* of A and denoted by A^o.

PROOF.– This is similar to the proof of proposition 2.7. □

EXERCISE 2.33.– Let A be a process with integrable variation in the wide sense and H a measurable bounded process. Show that, up to indistinguishability,

if H is an optional process, then $(H \cdot A)^o = H \cdot A^o$;

if H is a predictable process, then $(H \cdot A)^\pi = H \cdot A^\pi$;

if A is an optional process, then $(H \cdot A)^o = (OH) \cdot A$;

if A is a predictable process, then $(H \cdot A)^\pi = (\Pi H) \cdot A$.

HINT.– Use exercise 2.35 below.

EXERCISE 2.34.– Let A be a process with integrable variation in the wide sense. Show that, for every bounded random variable ξ and for every $t \in \mathbb{R}_+$,

$$\mathsf{E}\xi A_t^o = \lim_{n\to\infty} \mathsf{E}\left[\xi \sum_{k=1}^n \mathsf{E}\left(A_{\frac{k}{n}t} - A_{\frac{k-1}{n}t}\Big|\mathscr{F}_{\frac{k}{n}t}\right)\right],$$

$$\mathsf{E}\xi A_t^\pi = \lim_{n\to\infty} \mathsf{E}\left[\xi \sum_{k=1}^n \mathsf{E}\left(A_{\frac{k}{n}t} - A_{\frac{k-1}{n}t}\Big|\mathscr{F}_{\frac{k-1}{n}t}\right)\right].$$

EXERCISE 2.35.– Let A be a process with integrable variation in the wide sense, H a measurable process and $\mathsf{E}(|H| \cdot \mathrm{Var}\,(A)_\infty) < \infty$. Show that, μ_A-a.s.,

$$\frac{d\mu_{H\cdot A}}{d\mu_A} = H.$$

THEOREM 2.20.– Let A, $B \in \mathscr{V}$ (respectively A, $B \in \mathscr{V}^+$) and $dB.(\omega) \ll dA.(\omega)$ for almost all ω. Then there exists an optional (respectively a nonnegative optional) process H such that the processes B and $H \cdot A$ are indistinguishable. If A and B are predictable, then the process H can be chosen predictable.

PROOF.– We first prove the theorem under the assumption that A, $B \in \mathscr{A}^+$. Let C be a measurable set such that $\mu_A(C) = 0$. Then $\mathbb{1}_C \cdot A_\infty = 0$ a.s. It follows from the conditions of the theorem that $\mathbb{1}_C \cdot B_\infty = 0$ a.s. Hence, $\mu_B(C) = 0$ and, therefore, $\mu_B \ll \mu_A$. A fortiori, the absolute continuity persists if we consider the restrictions of these measures on the optional or predictable σ-algebra.

By the Radon–Nikodým theorem applied to the space $(\Omega \times \mathbb{R}_+, \mathscr{O})$ and the measures $\mu_A|_{\mathscr{O}}$ and $\mu_B|_{\mathscr{O}}$, there exists an optional process H such that

$\mu_B(C) = \int_C H \, d\mu_A$ for every $C \in \mathcal{O}$. Combining this relation with Exercise 2.35, we obtain that the Doléans measures of the processes $H \cdot A$ and B coincide on the σ-algebra $\mathcal{O}$. By theorem 2.14, the processes $H \cdot A$ and B are indistinguishable.

If, additionally, A and B are predictable, then one should modify the previous reasoning. Namely, we apply the Radon–Nikodým theorem to the measurable space $(\Omega \times \mathbb{R}_+, \mathscr{P})$ and the measures $\mu_A|_{\mathscr{P}}$, $\mu_B|_{\mathscr{P}}$, which gives us a predictable process H such that $\mu_B(C) = \int_C H \, d\mu_A$ for every $C \in \mathscr{P}$. Exercise 2.35 gives now that the Doléans measures of the processes $H \cdot A$ and B coincide on the σ-algebra $\mathscr{P}$, and, by theorem 2.15, the processes $H \cdot A$ and B are indistinguishable.

Now let $A, B \in \mathscr{V}^+$. Put $T_0 := 0$, $T_n := \inf \{t : A_t + B_t \geq n\}$. By proposition 1.10 (2), T_n is a stopping time and, moreover, it is predictable if A and B are predictable, see exercise 1.19. It is also clear that $T_n \uparrow +\infty$. Define the processes

$$A^n := \mathbb{1}_{[\![T_{n-1}, T_n[\![} \cdot A, \qquad B^n = \mathbb{1}_{[\![T_{n-1}, T_n[\![} \cdot B.$$

Since $A^n + B^n \leq n$, we have $A^n, B^n \in \mathscr{A}^+$; if A and B are predictable, then A^n and B^n are predictable. It is also obvious that $dB^n(\omega) \ll dA^n(\omega)$ for almost all ω. The first part of the proof yields nonnegative optional (predictable if A and B are predictable) processes H^n such that B^n and $H^n \cdot A^n$ are indistinguishable. Then the process $H := \sum_n H^n \mathbb{1}_{[\![T_{n-1}, T_n[\![}$ is nonnegative, optional (predictable, if A and B are predictable) and $B = H \cdot A$ up to an evanescent set.

The case, where $A \in \mathscr{V}^+$ and $B \in \mathscr{V}$, can be easily reduced to the considered case due to proposition 2.3.

Finally, let us consider the general case, where $A, B \in \mathscr{V}$. According to what has been proved, there is an optional (a predictable if A is predictable) process K such that $A = K \cdot \mathrm{Var}\,(A)$ up to indistinguishability. Then $\mathrm{Var}\,(A) = |K| \cdot \mathrm{Var}\,(A)$ up to indistinguishability, hence, $\mathbb{1}_{\{|K| \neq 1\}} \cdot \mathrm{Var}\,(A)_\infty = 0$ a.s. Put $J = K \mathbb{1}_{\{|K| = 1\}} + \mathbb{1}_{\{|K| \neq 1\}}$. The process J takes values ± 1, is optional (predictable if A is predictable), and it follows from the previous relation that the processes $K \cdot \mathrm{Var}\,(A)$ and $J \cdot \mathrm{Var}\,(A)$ are indistinguishable. Therefore, $A = J \cdot \mathrm{Var}\,(A)$ up to indistinguishability.

Using what has been proved in the previous case, there exists an optional (a predictable if A and B are predictable) process H' such that $B = H' \cdot \mathrm{Var}\,(A)$ up to indistinguishability. It remains to put $H := H'/J$. $\square$

2.5. Locally integrable increasing processes and processes with locally integrable variation

Unless otherwise stated, we will assume that a stochastic basis $\mathbb{B} = (\Omega, \mathscr{F}, \mathbb{F} = (\mathscr{F}_t)_{t \in \mathbb{R}_+}, \mathsf{P})$ satisfying the usual conditions is given.

DEFINITION 2.16.– A process with finite variation $A = (A_t)_{t \in \mathbb{R}_+} \in \mathscr{V}$ is called a *process with locally integrable variation* if there is a localizing sequence $\{T_n\}$ of stopping times such that $A^{T_n} \in \mathscr{A}$ for every n, i.e. the stopped process A^{T_n} is a process with integrable variation. An increasing process with locally integrable variation is called a *locally integrable increasing process*. The class of all processes with locally integrable variation is denoted by $\mathscr{A}_{\text{loc}}$, and the class of all locally integrable increasing processes is denoted by $\mathscr{A}_{\text{loc}}^+$.

It is clear that

$$\mathscr{A} \subseteq \mathscr{A}_{\text{loc}} \subseteq \mathscr{V}, \qquad \mathscr{A}^+ \subseteq \mathscr{A}_{\text{loc}}^+ = \mathscr{A}_{\text{loc}} \cap \mathscr{V}^+ \subseteq \mathscr{V}^+,$$

$$\mathscr{A}_{\text{loc}} = \mathscr{A}_{\text{loc}}^+ - \mathscr{A}_{\text{loc}}^+.$$

It is obvious from the definition that the process $A \in \mathscr{V}$ lies in $\mathscr{A}_{\text{loc}}$ if and only if $\text{Var}\,(A) \in \mathscr{A}_{\text{loc}}^+$, and that the classes $\mathscr{A}_{\text{loc}}$ and $\mathscr{A}_{\text{loc}}^+$ are stable under stopping. The first part of lemma 2.1 allows us to check trivially that $\mathscr{A}_{\text{loc}}$ is a linear space. "By localization" (i.e. applying the second part of lemma 2.1, see the proof of theorem 2.7), we can prove that $A \in \mathscr{A}_{\text{loc}}$ (respectively $A \in \mathscr{A}_{\text{loc}}^+$) if $A \in \mathscr{V}$ (respectively $A \in \mathscr{V}^+$) and there is a localizing sequence $\{T_n\}$ of stopping times such that $A^{T_n} \in \mathscr{A}_{\text{loc}}$ (respectively $A^{T_n} \in \mathscr{A}_{\text{loc}}^+$) for every n. Symbolically, $\mathscr{A}_{\text{loc}} = (\mathscr{A}_{\text{loc}})_{\text{loc}}$, $\mathscr{A}_{\text{loc}}^+ = (\mathscr{A}_{\text{loc}}^+)_{\text{loc}}$.

The following two important lemmas provide sufficient conditions for a process from $\mathscr{V}$ to be in $\mathscr{A}_{\text{loc}}$.

LEMMA 2.5.– Let $A \in \mathscr{V}$ and A be predictable. Then $A \in \mathscr{A}_{\text{loc}}$.

PROOF.– The process $B := \text{Var}\,(A)$ is predictable. Put $T_n := \inf\{t \colon B_t \geqslant n\}$. Then $T_n > 0$, $\lim_{n \to \infty} T_n = +\infty$, and, due to exercise 1.19, T_n is a predictable stopping time. Let $(S(n,p))_{p \in \mathbb{N}}$ be a foretelling sequence of stopping times for T_n. Choose p_n such that

$$\mathsf{P}(S(n, p_n) < T_n - 1) \leqslant 2^{-n},$$

and put $S_n = \max_{m \leqslant n} S(m, p_m)$. Then $S_n < \max_{m \leqslant n} T_m = T_n$, hence, $B_{S_n} \leqslant n$, hence $B^{S_n} \in \mathscr{A}^+$. Moreover, the sequence (S_n) is increasing and $\mathsf{P}(S_n < T_n - 1) \leqslant 2^{-n}$. Therefore, by the Borel–Cantelli lemma, for almost all ω, there is a number $n(\omega)$ such that $S_n(\omega) \geqslant T_n(\omega) - 1$ if $n \geqslant n(\omega)$. Thus, (S_n) is a localizing sequence. Consequently, $B \in \mathscr{A}_{\text{loc}}^+$. □

LEMMA 2.6.– Let $M \in \mathscr{V} \cap \mathscr{M}_{\text{loc}}$. Then $M \in \mathscr{A}_{\text{loc}}$.

PROOF.– Let (S_n) be a localizing sequence for M as an element of $\mathcal{M}_{\text{loc}}$, i.e. $M^{S_n} \in \mathcal{M}$ for every n. Put $T_n := S_n \wedge \inf\{t: B_t \geqslant n\}$, where $B := \text{Var}(M)$. It is clear that $T_n \uparrow \infty$ a.s. We have, for every $t \in \mathbb{R}_+$,

$$\Delta B_t = |\Delta M_t| \leqslant |M_t| + |M_{t-}| \qquad \text{and} \qquad |M_{t-}| \leqslant B_{t-}.$$

Therefore,

$$B_{T_n}\mathbb{1}_{\{T_n < \infty\}} = B_{T_n-}\mathbb{1}_{\{T_n < \infty\}} + \Delta B_{T_n}\mathbb{1}_{\{T_n < \infty\}}$$
$$\leqslant 2B_{T_n-}\mathbb{1}_{\{T_n < \infty\}} + |M_{T_n}|\mathbb{1}_{\{T_n < \infty\}} \leqslant 2n + |M_{T_n}|\mathbb{1}_{\{T_n < \infty\}},$$

hence $B_{T_n} \leqslant 2n + |M_{T_n}|$. It remains to note that the random variable $M_{T_n} = M^{S_n}_{T_n}$ is integrable by theorem 2.4. □

Before we turn to the main result of this section, we formulate and prove a technical assertion which will be used several times in the rest of the book.

Let a localizing sequence $\{T_n\}$ of stopping times and a sequence $\{X^n\}$ of measurable stochastic processes be given. Define S_n as the restriction of T_n on the set $A := \{\omega: \lim_n T_n(\omega) = \infty\}$; since $\mathsf{P}(A) = 1$, we have $A \in \mathscr{F}_{T_n}$ for every n, hence, S_n is a stopping time by proposition 1.5. Clearly

$$S_1(\omega) \leqslant S_2(\omega) \leqslant \ldots \leqslant S_n(\omega) \leqslant \ldots \qquad \text{and} \qquad \lim_n S_n(\omega) = \infty.$$

for all ω. Now define the process X recursively, putting, for all $t \in \mathbb{R}_+$ and ω,

$$X_t(\omega) = \begin{cases} X^1_t(\omega), \\ \qquad \text{if } t \in [0, S_1(\omega)]; \\ X_{S_{n-1}(\omega)}(\omega) + \left(X^n_t(\omega) - X^n_{S_{n-1}(\omega)}(\omega)\right), \\ \qquad \text{if } t \in]S_{n-1}(\omega), S_n(\omega)], n = 2, 3, \ldots. \end{cases} \qquad [2.34]$$

It is easy to see that, equivalently, X can be written as:

$$X = X^1_0 + \sum_{n=1}^{\infty} \left[(X^n)^{S_n} - (X^n)^{S_{n-1}}\right], \qquad [2.35]$$

where $S_0 = 0$. Only finite number of terms do not equal zero in this sum for every ω and t, namely, corresponding to n such that $S_{n-1}(\omega) < t$. Hence, X is a measurable stochastic process. We will say that X given by [2.34] or [2.35] is obtained by the "gluing" procedure from $\{X^n\}, \{T_n\}$.

Some properties of "gluing" are described in the following proposition.

PROPOSITION 2.9.– Let X be obtained by the "gluing" procedure from $\{X^n\}$, $\{T_n\}$, where $\{T_n\}$ is a localizing sequence of stopping times and $\{X^n\}$ is a sequence of measurable processes. Then:

1) if all X^n are càdlàg (respectively continuous), then X is càdlàg (respectively continuous);

2) if all X^n are progressively measurable (respectively optional, respectively predictable), then X is progressively measurable (respectively optional, respectively predictable);

3) if $X^n \in \mathcal{M}_{\mathrm{loc}}$ for all n, then $X \in \mathcal{M}_{\mathrm{loc}}$;

4) if $X^n \in \mathcal{V}$ (respectively $X^n \in \mathcal{V}^+$) for all n, then $X \in \mathcal{V}$ (respectively $X \in \mathcal{V}^+$);

5) if $(X^n)^{T_{n-1}}$ and $(X^{n-1})^{T_{n-1}}$ are indistinguishable for every $n = 2, 3, \ldots$, then X^{T_n} and $(X^n)^{T_n}$ are indistinguishable for every $n = 1, 2, \ldots$.

PROOF.–

The proof (1) is obvious from [2.34], and (2) follows trivially from [2.35] and proposition 1.9. It follows from [2.35] that $X^{S_k} = X_0^1 + \sum_{n=1}^k \left[(X^n)^{S_n} - (X^n)^{S_{n-1}} \right]$, hence, under assumption (3), $X^{S_k} \in \mathcal{M}_{\mathrm{loc}}$ for every k, and the claim follows from theorem 2.7. Assertion (4) is evident in view of [2.34]. Let assumptions in (5) be satisfied. Then $X^n(\omega)$ and $X^{n-1}(\omega)$ coincide on $[0, S_{n-1}(\omega)]$. By induction on n, it is proved from [2.34] that, for $\omega \in B$, $X_t(\omega) = X_t^n(\omega)$ for all $t \in [0, S_n(\omega)]$. Therefore, X^{S_n} and $(X^n)^{S_n}$ are indistinguishable, hence X^{T_n} and $(X^n)^{T_n}$ are indistinguishable. □

The next theorem is the main result of this section.

THEOREM 2.21.–

1) Let $A \in \mathcal{V}$. The following two statements are equivalent:

a) $A \in \mathcal{A}_{\mathrm{loc}}$;

b) there exists a predictable process $\widetilde{A} \in \mathcal{V}$ such that $A - \widetilde{A} \in \mathcal{M}_{\mathrm{loc}}$.

Such a process $\widetilde{A}$ (if it exists) is unique up to indistinguishability and called the compensator of A. If $A \in \mathcal{A}_{\mathrm{loc}}^+$, then the compensator $\widetilde{A}$ can be chosen in $\mathcal{V}^+$.

2) Let $A \in \mathcal{A}_{\mathrm{loc}}$, $\widetilde{A}$ being the compensator of A, $H \in L_{\mathrm{var}}(A)$, and $H \cdot A \in \mathcal{A}_{\mathrm{loc}}$. Then $H \in L_{\mathrm{var}}(\widetilde{A})$, $H \cdot \widetilde{A} \in \mathcal{A}_{\mathrm{loc}}$, and $H \cdot A - H \cdot \widetilde{A} \in \mathcal{M}_{\mathrm{loc}}$.

3) Let $A \in \mathcal{A}_{\mathrm{loc}}^+$, $B \in \mathcal{V}^+$, and let B be predictable. The following statements are equivalent:

c) B is a compensator of A;

d) $\mathsf{E}A_T = \mathsf{E}B_T$ for every stopping time T;

e) $\mathsf{E} \displaystyle\int_0^\infty H_t \, dA_t = \mathsf{E} \int_0^\infty H_t \, dB_t$

for every nonnegative predictable process H.

REMARK 2.9.– If A is a process with integrable variation, the definition of its compensator given in theorem 2.21 coincides with the previous definition in section 2.4.

REMARK 2.10.– As example 2.3 below shows, the assumption $H \cdot A \in \mathscr{A}_{\mathrm{loc}}$ in assertion (2) is essential. In general, it cannot be replaced by the assumption $H \in L_{\mathrm{var}}(\widetilde{A})$ alone or combined with $H \cdot \widetilde{A} \in \mathscr{A}_{\mathrm{loc}}$. Nevertheless, it should be noted that, as follows from assertion (3) of the theorem, for $A \in \mathscr{V}^+$, the assumptions $H \in L_{\mathrm{var}}(\widetilde{A})$ and $H \cdot \widetilde{A} \in \mathscr{A}_{\mathrm{loc}}$ imply $H \in L_{\mathrm{var}}(A)$ and $H \cdot A \in \mathscr{A}_{\mathrm{loc}}$.

It follows from theorems 2.21 and 2.10 that lemma 2.4 and theorem 2.18 are valid for locally integrable increasing processes.

COROLLARY 2.12.– If $A \in \mathscr{A}_{\mathrm{loc}}$, then, for every predictable stopping time T, a.s.,

$$\mathsf{E}\big(|\Delta A_T|\mathbb{1}_{\{T<\infty\}}\big|\mathscr{F}_{T-}\big) < \infty$$

and

$$\mathsf{E}\big(\Delta A_T \mathbb{1}_{\{T<\infty\}}\big|\mathscr{F}_{T-}\big) = \Delta \widetilde{A}_T \mathbb{1}_{\{T<\infty\}},$$

where $\widetilde{A}$ is a compensator of A.

COROLLARY 2.13.– Let $A \in \mathscr{A}_{\mathrm{loc}}^+$. The compensator $\widetilde{A}$ of A is a.s. continuous if and only if $\Delta A_T \mathbb{1}_{\{T<\infty\}} = 0$ a.s. for every predictable stopping time T.

EXAMPLE 2.3 (Émery).– Let S and η be independent random variables on a complete probability space. It is assumed that $S(\omega) > 0$ for all ω and S has the exponential distribution: $\mathsf{P}(S > t) = e^{-t}$, $t \in \mathbb{R}_+$, while $\mathsf{P}(\eta = \pm 1) = 1/2$. Put

$$X_t = \begin{cases} 0, & \text{if } t < S, \\ \eta/S, & \text{if } t \geqslant S, \end{cases} \quad A_t = \begin{cases} 0, & \text{if } t < S, \\ \eta, & \text{if } t \geqslant S, \end{cases} \quad H_t = \frac{1}{t}\mathbb{1}_{\{t>0\}}.$$

Put $\mathscr{F}_t^0 = \sigma\{X_s,\ s \leqslant t\}$, $\mathbb{F}^0 = (\mathscr{F}_t^0)_{t \in \mathbb{R}_+}$, and define a filtration $\mathbb{F}$ on $(\Omega, \mathscr{F}, \mathsf{P})$ by $\mathbb{F} := (\mathbb{F}^{0+})^{\mathsf{P}}$; see exercises 1.2–1.4. Then $A \in \mathscr{A}_{\mathrm{loc}} \cap \mathscr{M}_{\mathrm{loc}}$, $H \in L_{\mathrm{var}}(A)$ and $H \cdot A = X \in \mathscr{V}$, but $X \notin \mathscr{A}_{\mathrm{loc}}$ and $X \notin \mathscr{M}_{\mathrm{loc}}$.

EXERCISE 2.36.– Show that $A \in \mathcal{M}_{loc}$, $X \notin \mathcal{A}_{loc}$, and $X \notin \mathcal{M}_{loc}$ in example 2.3.

HINT.– Show that $E|X_T| = \infty$ for every stopping time T with $P(T > 0) > 0$. To do so, find a characterization of stopping times, similar to that in exercise 1.32.

PROOF OF THEOREM 2.21.– (1) If (b) holds, then, by lemmas 2.5 and 2.6, $\widetilde{A} \in \mathcal{A}_{loc}$ and $A - \widetilde{A} \in \mathcal{A}_{loc}$. Therefore, $A \in \mathcal{A}_{loc}$.

Let $A \in \mathcal{A}_{loc}$. Take a localizing sequence (T_n) such that $A^{T_n} \in \mathcal{A}$ for all n. By corollary 2.11, there exists a compensator of $A^n := A^{T_n}$, i.e. a predictable process $\widetilde{A}^n \in \mathcal{A}$ such that $A^n - \widetilde{A}^n \in \mathcal{M}$ (moreover, $\widetilde{A}^n \in \mathcal{A}^+$ if $A \in \mathcal{V}^+$). Note that $\left(A^{T_n}\right)^{T_{n-1}} = A^{T_{n-1}}$. This implies, in view of the uniqueness in corollary 2.11, that $\widetilde{A}^n = \left(\widetilde{A}^n\right)^{T_n}$ and $\left(\widetilde{A}^n\right)^{T_{n-1}} = \widetilde{A}^{n-1} = \left(\widetilde{A}^{n-1}\right)^{T_{n-1}}$ up to indistinguishability.

Let $\widetilde{A}$ be obtained from $\{\widetilde{A}^n\}$, $\{T_n\}$ by the "gluing" procedure. Owing to proposition 2.9, we have: $\widetilde{A} \in \mathcal{V}$, $\widetilde{A}$ is predictable, $\widetilde{A} \in \mathcal{V}^+$ if $A \in \mathcal{V}^+$, and $\widetilde{A}^{T_n}$ is indistinguishable with $\widetilde{A}^n$, hence $A^{T_n} - \widetilde{A}^{T_n} \in \mathcal{M}$. Thus, $A - \widetilde{A} \in \mathcal{M}_{loc}$.

It remains to prove the uniqueness of the compensator. If $\widetilde{A}$ and $\widetilde{A}'$ satisfy the definition of the compensator, then the process $B := \widetilde{A} - \widetilde{A}' \in \mathcal{V} \cap \mathcal{M}_{loc}$ and is predictable. By lemma 2.5 (or 2.6), there is a localizing sequence (T_n) such that $B^{T_n} \in \mathcal{A}$. By corollary 2.10, B^{T_n} is indistinguishable with the zero process for every n.

(3) Assertion (d) is a special case of assertion (e) with $H = \mathbb{1}_{[0,T]}$, so implication (e)$\Rightarrow$(d) is obvious.

Let us prove (d)$\Rightarrow$(c). Put $M := A - B$. Let a localizing sequence (T_n) be such that $A^{T_n} \in \mathcal{A}^+$, then $B^{T_n} \in \mathcal{A}^+$ in view of (d). Hence, $M^{T_n} \in \mathcal{A}$; moreover, for every stopping time T,

$$EM_T^{T_n} = E(A_{T_n \wedge T} - B_{T_n \wedge T}) = 0$$

because of (d). Write this relation for $T = s\mathbb{1}_B + t\mathbb{1}_{\Omega \setminus B}$ and $T = t$, where $s < t$ and $B \in \mathcal{F}_s$, then subtract the obtained relations from each other. As a result, we get $E(M_t^{T_n} - M_s^{T_n})\mathbb{1}_B = 0$, hence, $M^{T_n} \in \mathcal{M}$ and $M \in \mathcal{M}_{loc}$.

It remains to check implication (c)$\Rightarrow$(e). Let us take a common localizing sequence (T_n) for $A \in \mathcal{A}_{loc}$ and $B \in \mathcal{A}_{loc}$ (which is possible by lemma 2.1 (1)). Then $M^{T_n} \in \mathcal{M}$ for every n because of (c), where M is defined as above. Therefore, by theorem 2.13, the Doléans measures of the processes A^{T_n} and B^{T_n} coincide on the σ-algebra $\mathscr{P}$. Hence, integrals of predictable functions coincide as well: for nonnegative predictable H,

$$\mathsf{E} \int_0^\infty H_s \mathbb{1}_{\{H_s \leqslant n\}} \mathbb{1}_{[\![0,T_n]\!]} \, dA_s = \int_{\Omega \times \mathbb{R}_+} H \mathbb{1}_{\{H \leqslant n\}} \, d\mu_{A^{T_n}}$$

$$= \int_{\Omega \times \mathbb{R}_+} H_s \mathbb{1}_{\{H_s \leqslant n\}} \, d\mu_{B^{T_n}} = \mathsf{E} \int_0^\infty H \mathbb{1}_{\{H \leqslant n\}} \mathbb{1}_{[\![0,T_n]\!]} \, dB_s.$$

It remains to pass to the limit as $n \to \infty$ in terms on the left and on the right in the above relation. We use the theorem on monotone convergence, first, for the inner Lebesgue–Stieltjes integrals, then for the external integrals with respect to P.

(2) Let $A = B - C$ be the decomposition of A from proposition 2.3. Since

$$\mathrm{Var}\,(H \cdot A) = |H| \cdot \mathrm{Var}\,(A) = H^+ \cdot B + H^+ \cdot C + H^- \cdot B + H^- \cdot C,$$

each of the processes $H^+ \cdot B$, $H^+ \cdot C$, $H^- \cdot B$, $H^- \cdot C$ is locally integrable and, hence, has a compensator. Due to implication (c)$\Rightarrow$(e) from assertion (3), we have, for every stopping time T,

$$\mathsf{E} \int_0^\infty H_s^+ \mathbb{1}_{[\![0,T]\!]} \, dB_s = \mathsf{E} \int_0^\infty H_s^+ \mathbb{1}_{[\![0,T]\!]} \, d\widetilde{B}_s,$$

where $\widetilde{B}$ is the compensator of B. The left side of this equality is finite for $T = T_n$, where (T_n) is a localizing sequence for $H^+ \cdot B$, and it follows that $H^+ \in L_{\mathrm{var}}(\widetilde{B})$. Moreover, $H^+ \cdot \widetilde{B}$ is the compensator of $H^+ \cdot B$ by implication (d)$\Rightarrow$(c) from assertion (3), i.e. $H^+ \cdot B - H^+ \cdot \widetilde{B} \in \mathscr{M}_{\mathrm{loc}}$. Applying the same arguments to processes $H^+ \cdot C$, $H^- \cdot B$, $H^- \cdot C$, we get $H \cdot A - H \cdot \widetilde{A} \in \mathscr{M}_{\mathrm{loc}}$.

The theorem has been proved. $\square$

For ease of reference, we state an assertion equivalent to the uniqueness of the compensator.

COROLLARY 2.14.– A predictable local martingale with finite variation is indistinguishable with the zero process.

Some properties of the compensator are formulated in the next proposition. A tilde over a process signifies its compensator.

PROPOSITION 2.10.–

1) If $A \in \mathcal{V}$ is predictable, then $\widetilde{A} = A$.

2) If $A \in \mathcal{A}_{\mathrm{loc}}$ and T is a stopping time, then $\widetilde{A^T} = \widetilde{A}^T$.

3) If $A \in \mathcal{A}_{\mathrm{loc}}$, then $A \in \mathcal{M}_{\mathrm{loc}}$ if and only if $\widetilde{A} = 0$.

4) If $M \in \mathcal{M}_{\mathrm{loc}} \cap \mathcal{V}$, $H \in L_{\mathrm{var}}(M)$, and $H \cdot M \in \mathcal{A}_{\mathrm{loc}}$, then $H \cdot M \in \mathcal{M}_{\mathrm{loc}}$. In particular, if $M \in \mathcal{M}_{\mathrm{loc}} \cap \mathcal{V}$ and H is a locally bounded predictable process, then $H \cdot M \in \mathcal{M}_{\mathrm{loc}}$.

5) If $M \in \mathcal{M}_{\mathrm{loc}} \cap \mathcal{V}$ and $A_t = \sum_{0 < s \leqslant t} \Delta M_s$, then $A = (A_t) \in \mathcal{A}_{\mathrm{loc}}$, the compensator $\widetilde{A}$ has a continuous version, and $M = A - \widetilde{A}$.

Equalities in (1)–(3) are assumed up to an evanescent set.

Assertion (4) says that the Lebesgue–Stieltjes integral $H \cdot M$ of a predictable function H with respect to a local martingale M with finite variation is a local martingale but only under the assumption that $H \cdot M \in \mathcal{A}_{\mathrm{loc}}$; see example 2.3.

Assertion (5) means that a local martingale with finite variation is the compensated sum of its jumps.

PROOF.– Assertions (1)–(3) are obvious, and (4) follows from theorem 2.21 (2). Let us prove (5). Since $\mathrm{Var}\,(A) \leqslant \mathrm{Var}\,(M)$, we have $A \in \mathcal{A}_{\mathrm{loc}}$. Next, the process $A - M$ belongs to $\mathcal{V}$ and is continuous, in particular, $A - M$ is predictable. Since $A - (A - M) = M$, it follows from the definition of compensator that $A - M$ is the compensator of A.   □

LEMMA 2.7.– Let $M \in \mathcal{M}_{\mathrm{loc}}$. Then $(M - M_0)^* \in \mathcal{A}_{\mathrm{loc}}^+$ and $(\Delta M)^* \in \mathcal{A}_{\mathrm{loc}}^+$.

PROOF.– It is evident that trajectories of $(M - M_0)^*$ and $(\Delta M)^*$ are monotone, and their right-continuity follows from right-continuity of M. Both processes are adapted because M is adapted and right-continuous. By theorem 2.7, there is a localizing sequence (T_n) such that $M^{T_n} \in \mathcal{H}^1$, i.e. $\mathbb{E} M_{T_n}^* < \infty$ for every n. It remains to note that $(M - M_0)^* \leqslant 2M^*$ and $(\Delta M)^* \leqslant 2M^*$.   □

THEOREM 2.22 (the Gundy decomposition).– Every local martingale M has a decomposition $M = M^1 + M^2$, where M^1 and M^2 are local martingales, jumps ΔM^1 are uniformly bounded, and $M^2 \in \mathcal{V}$.

PROOF.– Put

$$A_t = \sum_{0 < s \leqslant t} \Delta M_s \mathbb{1}_{\{|\Delta M_s| > 1\}}.$$

It is clear that $A = (A_t) \in \mathcal{V}$. Put $T_n := \inf\{t\colon \operatorname{Var}(A)_t \geqslant n\}$, then (T_n) is a localizing sequence. We have $\operatorname{Var}(A)^{T_n} \leqslant n + (\Delta M)^*$, where $\operatorname{Var}(A)^{T_n} \in \mathscr{A}_{\mathrm{loc}}^+$ by lemma 2.7. Therefore, $A \in \mathscr{A}_{\mathrm{loc}}$.

Now define $M^2 := A - \widetilde{A}$, where $\widetilde{A}$ is the compensator of A, and let $M^1 := M - M^2$. Since $\Delta M^1 = \Delta M - \Delta A + \Delta \widetilde{A} = \Delta M \mathbb{1}_{\{|\Delta M| \leqslant 1\}} + \Delta \widetilde{A}$, it remains to prove that absolute values of jumps $\Delta \widetilde{A}$ are bounded by one.

Note that $|\Delta \widetilde{A}_T| \mathbb{1}_{\{T < \infty\}} \leqslant 1$ a.s. for every predictable stopping time T. Indeed, this follows from corollary 2.12 and theorem 2.10:

$$\Delta \widetilde{A}_T \mathbb{1}_{\{T<\infty\}} = \mathsf{E}\big(\Delta A_T \mathbb{1}_{\{T<\infty\}} \big| \mathscr{F}_{T-}\big)$$
$$= \mathsf{E}\big(\Delta M_T \mathbb{1}_{\{T<\infty\}} \big| \mathscr{F}_{T-}\big) - \mathsf{E}\big(\Delta M_T \mathbb{1}_{\{|\Delta M_T| \leqslant 1\}} \mathbb{1}_{\{T<\infty\}} \big| \mathscr{F}_{T-}\big)$$
$$= -\mathsf{E}\big(\Delta M_T \mathbb{1}_{\{|\Delta M_T| \leqslant 1\}} \mathbb{1}_{\{T<\infty\}} \big| \mathscr{F}_{T-}\big).$$

We can conclude from theorem 1.18 that the set $\{|\Delta \widetilde{A}_T| > 1\}$ is evanescent. Now, replacing $\widetilde{A}$ by the process $\widetilde{A} - \sum_{0<s\leqslant\cdot} \Delta \widetilde{A} \mathbb{1}_{\{|\Delta \widetilde{A}| > 1\}}$, which is indistinguishable from $\widetilde{A}$, we obtain a version of $\widetilde{A}$ such that $|\Delta \widetilde{A}| \leqslant 1$ identically. $\qquad\Box$

EXAMPLE 2.4.– Let us consider the model from exercises 1.27–1.35. Assume that $S(\omega) > 0$ for all ω and S has the exponential distribution: $\mathsf{P}(S > t) = e^{-t}$. Put $A := \mathbb{1}_{[\![S,\infty[\![}$. The compensator of the integrable increasing process A is the process $\widetilde{A}$ given by $\widetilde{A}_t := S \wedge t$, see exercise 2.37 below.

Let M be a nonnegative uniformly integrable martingale on the considered stochastic basis. Assume first that M is bounded. Then

$$\mathsf{E}M_{S-} = \mathsf{E}(M_- \cdot A)_\infty = \mathsf{E}(M_- \cdot \widetilde{A})_\infty \overset{(*)}{=} \mathsf{E}M_\infty \widetilde{A}_\infty = \mathsf{E}M_\infty S,$$

where the equality marked by $(*)$ follows from lemma 2.3. (In fact, the formula above is valid without the assumption that M is bounded; however, it needs to be justified.)

Now let the martingale M be defined through its terminal value M_∞ given by

$$M_\infty := S^{-2} e^S \mathbb{1}_{\{S \geqslant 1\}}.$$

Also define nonnegative martingales M^n by $M_\infty^n := M_\infty \mathbb{1}_{\{S \leqslant n\}}$. Using what has been proved, we get

$$\mathsf{E}M_{S-} \geqslant \mathsf{E}M_{S-}^n = \mathsf{E}M_\infty^n S = \mathsf{E}S^{-1} e^S \mathbb{1}_{\{1 \leqslant S \leqslant n\}}$$
$$= \int_1^n x^{-1} e^x e^{-x} \, dx = \log n.$$

Therefore,

$$\mathsf{E} M_{S-} = \infty.$$

In particular, $M \in \mathscr{M}$ but $M \notin \mathscr{H}^1$.

EXERCISE 2.37.– Show that $\widetilde{A}$ is the compensator of A in Example 2.4.

HINT.– Use implication (d)$\Rightarrow$(c) from theorem 2.21 (3) and exercise 1.32.

EXERCISE 2.38.– Let $X = (X_t)_{t \in \mathbb{R}_+}$ be a Poisson process with intensity 1 on a stochastic basis $(\Omega, \mathscr{F}, \mathbb{F}, \mathsf{P})$. Put $S := \inf \{t \colon X_t \geqslant 1\}$. Find the compensators of X and X^S and compare answers with exercise 2.37.

HINT.– The compensator of X is a deterministic function.

EXERCISE 2.39.– As in example 2.4, consider the model from exercises 1.27–1.35, but here do not assume that S has the exponential distribution. Put $F(t) := \mathsf{P}(S \leqslant t)$, $A := \mathbb{1}_{[\![S,\infty[\![}$,

$$\widetilde{A}_t := \int\limits_{0}^{S \wedge t} \frac{dF(s)}{1 - F(s - 0)}.$$

Show that the process $\widetilde{A} = (\widetilde{A}_t)$ is the compensator of A. Show that, if the distribution of S is continuous, then

$$\widetilde{A}_t = \log \frac{1}{1 - F(S \wedge t)}.$$

Let T be a strictly positive totally inaccessible or predictable stopping time. In the next two lemmas, we provide sufficient conditions on a random variable ξ in order for there to exist a local martingale M with the jump ξ at T; more precisely, $\xi = \Delta M_T \mathbb{1}_{\{T < \infty\}}$. In fact, conditions that we suggest are also necessary: this follows from lemma 2.7 for totally inaccessible T and from theorem 2.10 for predictable T.

LEMMA 2.8.– Let T be a strictly positive totally inaccessible stopping time and ξ an $\mathscr{F}_T$-measurable random variable vanishing on $\{T = \infty\}$. Assume that the process $A := \xi \mathbb{1}_{[\![T,\infty[\![}$ is a process with locally integrable variation. Then the compensator $\widetilde{A}$ of A has a continuous version.

PROOF.– If $\xi \geqslant 0$, the claim follows from corollary 2.13. The general case follows from the decomposition $\xi = \xi^+ - \xi^-$.   □

LEMMA 2.9.– Let T be a strictly positive predictable stopping time and ξ be an $\mathscr{F}_T$-measurable random variable vanishing on $\{T = \infty\}$, $\mathsf{E}(|\xi||\mathscr{F}_{T-}) < \infty$, $\mathsf{E}(\xi|\mathscr{F}_{T-}) = 0$. Put $A := \xi \mathbb{1}_{[\![T,\infty[\![}}$. Then $A \in \mathscr{A}_{\mathrm{loc}}$, and the compensator of A is zero.

PROOF.– We need to prove that $A \in \mathscr{M}_{\mathrm{loc}}$. Let us first assume that ξ is integrable and show that $A \in \mathscr{M}$. In other words, we want to show that, if M is a uniformly integrable martingale such that $M_\infty = \xi$, then M and A are indistinguishable.

In the proof of theorem 2.16, it was shown that if ξ is, additionally, bounded, then $M_0 = 0$ a.s. and the process $M\mathbb{1}_{[\![0,T[\![}}$ is indistinguishable with the zero process. It is easy to see that boundedness of ξ is not fully used, and the arguments remain true for integrable ξ. Next, ξ is $\mathscr{F}_T$-measurable, hence, the random variable $\xi\mathbb{1}_{\{T \leqslant t\}}$ is $\mathscr{F}_t$-measurable for every $t \in \mathbb{R}_+$. Therefore,

$$M_t \mathbb{1}_{\{T \leqslant t\}} = \mathsf{E}(\xi|\mathscr{F}_t)\mathbb{1}_{\{T \leqslant t\}} = \mathsf{E}(\xi\mathbb{1}_{\{T \leqslant t\}}|\mathscr{F}_t) = \xi\mathbb{1}_{\{T \leqslant t\}} = A_t.$$

Since $M\mathbb{1}_{[\![T,\infty[\![}}$ and A are right-continuous, they are indistinguishable. Thus, M and A are indistinguishable.

Now assume that a random variable ξ satisfies the assumptions of the lemma but is not integrable. By proposition A.2, there is an increasing sequence of sets B_n from $\mathscr{F}_{T-}$ such that $\cup_n B_n = \Omega$ a.s. and $\mathsf{E}|\xi|\mathbb{1}_{B_n} < \infty$. Put $T_n := T_{\Omega \setminus B_n}$, i.e. T_n is the restriction of T on the complement of B_n. It is clear that (T_n) is a localizing sequence; moreover, all T_n are predictable. As in the proof of lemma 2.5, we construct a localizing sequence (S_n) such that $S_n < T_n$ for every n. Also put $A^n := \xi\mathbb{1}_{B_n}\mathbb{1}_{[\![T,\infty[\![}}$. We have proved above that $A^n \in \mathscr{M}$. But $A^{S_n} = (A^n)^{S_n} \in \mathscr{M}$, hence, $A \in \mathscr{M}_{\mathrm{loc}}$. □

The following technical lemma will often be used below.

LEMMA 2.10.– Let L be an adapted càdlàg process. Assume that there is a finite limit $\lim_{t \to \infty} L_t =: L_\infty$ a.s., $\mathsf{E}|L_T| < \infty$ and $\mathsf{E}L_T = 0$ for every stopping time T. Then $L \in \mathscr{M}$.

PROOF.– Let $t \in \mathbb{R}_+$, $B \in \mathscr{F}_t$. For $T = t_B$ and $T = \infty$, the equality $\mathsf{E}L_T = 0$ is written as $\mathsf{E}(L_t \mathbb{1}_B + L_\infty \mathbb{1}_{\Omega \setminus B}) = 0$ and $\mathsf{E}L_\infty = 0$, respectively. Hence, $\mathsf{E}L_t \mathbb{1}_B = \mathsf{E}L_\infty \mathbb{1}_B$, i.e., $L_t = \mathsf{E}(L_\infty|\mathscr{F}_t)$ a.s. for every $t \in \mathbb{R}_+$. □

LEMMA 2.11.– Let $M \in \mathscr{M}_{\mathrm{loc}} \cap \mathscr{A}$ and N be a bounded martingale. Then

$$\mathsf{E}M_\infty N_\infty = \mathsf{E}\sum_{s \in \mathbb{R}} \Delta M_s \Delta N_s \qquad [2.36]$$

and the process $L = (L_t)$, where

$$L_t := M_t N_t - \sum_{0 < s \leqslant t} \Delta M_s \Delta N_s,$$

is a uniformly integrable martingale.

PROOF.– By lemma 2.3,

$$\mathsf{E} M_\infty N_\infty = \mathsf{E}(N \cdot M_\infty).$$

Moreover,

$$\mathsf{E}(N_- \cdot M_\infty) = 0,$$

because N_- is predictable and the restriction of the Doléans measure μ_M onto the predictable σ-algebra is zero. Hence,

$$\mathsf{E} M_\infty N_\infty = \mathsf{E}((\Delta N) \cdot M_\infty) = \mathsf{E} \sum_{s \in \mathbb{R}} \Delta M_s \Delta N_s.$$

Now we can apply [2.36] to N^T instead of N, where T is an arbitrary stopping time:

$$\mathsf{E} M_T N_T = \mathsf{E} M_\infty N_T = \mathsf{E} \sum_{s \leqslant T} \Delta M_s \Delta N_s,$$

i.e. $\mathsf{E} L_T = 0$. So the second statement of the lemma follows from lemma 2.10.   □

2.6. Doob–Meyer decomposition

We will assume that a stochastic basis $\mathbb{B} = (\Omega, \mathscr{F}, \mathbb{F} = (\mathscr{F}_t)_{t \in \mathbb{R}_+}, \mathsf{P})$ satisfying the usual conditions is given.

In the discrete-time case, it is easy to prove that every submartingale can be decomposed into the sum of a martingale and a predictable increasing process. This decomposition is unique and called the Doob decomposition.

In the continuous-time case, such a decomposition may not exist. Indeed, we considered examples 2.1 and 2.2 of local martingales, which are supermartingales and not martingales. Changing the sign, take a submartingale X which is a local martingale and not a martingale. If there were a decomposition $X = M + A$ into a martingale M and a predictable increasing process A, then A would be a predictable local martingale with bounded variation. Hence, $A = 0$ by corollary 2.14, and we arrive at a contradiction. Moreover, a contradiction arises even if we do not require A to be predictable: $A \in \mathcal{M}_{\text{loc}} \cap \mathcal{V}^+$ implies $A = 0$; see corollary 2.9.

Necessary and sufficient conditions for the existence and uniqueness of a decomposition of a submartingale into the sum of a martingale and a natural (!) increasing process were obtained by P.-A. Meyer, so it is called the Doob–Meyer decomposition.

We begin with some preliminary considerations. If a process X can be represented as $X = M + A$, where $M \in \mathcal{M}_{\text{loc}}$, $A \in \mathcal{V}$ and A is predictable, then such a decomposition is unique (up to indistinguishability) by corollary 2.14. The same argument guarantees the uniqueness of the decomposition in all theorems below.

Let a submartingale X admit the decomposition $X = M + A$ into the sum of a martingale M and an increasing process A (the predictability of A is not assumed). Then $\mathsf{E}A_t = \mathsf{E}X_t - \mathsf{E}M_t < \infty$ for every $t \in \mathbb{R}_+$. Since the stopped process A^t is majorized by a random variable A_t, we have $A \in (DL)$. Since $M \in (DL)$, see corollary 2.4, we have $X \in (DL)$; see theorem A.5 (2). Thus, a necessary condition for the existence of the decomposition under consideration is that the submartingale X belongs to the class (DL). We will see that this condition is also sufficient for the existence of the decomposition even with a predictable A.

It is also easy to prove the necessity of another more technical condition for the existence of the decomposition under consideration. That condition also turns out to be sufficient, and the main difficulty in the proposed proof of the Doob–Meyer decomposition is to prove that this technical condition follows from the condition that X belongs to the class (DL). We first introduce some additional notation.

Denote by $\mathcal{R}$, the collection of all sets of the form $B \times \{0\}$, $B \in \mathcal{F}_0$, or $B \times]s, t]$, $s, t \in \mathbb{R}_+$, $s < t$, $B \in \mathcal{F}_s$. It is clear that $\mathcal{R}$ is a semi-ring. Recall that $\mathcal{P} = \sigma\{\mathcal{R}\}$ by theorem 1.10. It is well known that finite unions of pairwise disjoint sets from $\mathcal{R}$ form a ring that we denote by $\mathcal{J}$.

Let X be a stochastic process. For $C \in \mathcal{R}$, define a random variable $X(C)$ as $\mathbb{1}_B(X_t - X_s)$, if $C = B \times]s, t]$, $s, t \in \mathbb{R}_+$, $s < t$, $B \in \mathcal{F}_s$; if $C = B \times \{0\}$, $B \in \mathcal{F}_0$, we put $X(C) := 0$. If a set $J \in \mathcal{J}$ is represented as the union of pairwise disjoint sets $C_1, \dots, C_n$ from $\mathcal{R}$, put $X(J) := \sum_{k=1}^n X(C_k)$. This definition is correct; moreover, if a set $J \in \mathcal{J}$ is represented as the union of pairwise disjoint sets $J_1, \dots, J_n$ from $\mathcal{J}$, then $X(J) = \sum_{k=1}^n X(J_k)$.

Now let us assume that X is an adapted càdlàg process and $\mathsf{E}|X_t| < \infty$ for every $t \in \mathbb{R}_+$. Define a function $\mu = \mu_X$ on $\mathcal{J}$ by $\mu(J) := \mathsf{E}X(J)$. According to the previous paragraph, μ is an additive set function on the ring $\mathscr{J}$. It follows from the definition of a submartingale that X is a submartingale if and only if μ is nonnegative on $\mathscr{R}$ and, hence, on $\mathscr{J}$ (recall that we consider only right-continuous submartingales).

Now let a submartingale X admit a decomposition $X = M + A$ into the sum of a martingale M and an increasing process A. Then $\mathsf{E}A_t < \infty$ for every $t \in \mathbb{R}_+$. Obviously, $\mu = \mu_X$ and μ_A coincide on $\mathscr{R}$ and, hence, on $\mathscr{J}$. But μ_A can be extended to a σ-finite (countably additive) measure on $\mathscr{P} = \sigma\{\mathscr{J}\}$, namely, to the restriction of the Doléans measure of A onto $\mathscr{P}$. (In fact, the Doléans measure was defined for processes $A \in \mathscr{A}$. However, if $A \in \mathscr{A}_{\mathrm{loc}}^+$, we can give exactly the same definition. Then the Doléans measure takes values in $[0, +\infty]$ and its restriction onto $\mathscr{P}$ is a σ-finite measure.) Thus, for the existence of the above decomposition, it is necessary that the measure μ can be extended from the ring $\mathscr{J}$ to a countably additive measure on $\mathscr{P}$. The sufficiency of this condition is also easy to check: take as A the predictable increasing process such that the restriction of its Doléans measure onto $\mathscr{P}$ coincides with the extension of μ on $\mathscr{P}$, see theorem 2.17. We realize this idea below. For technical reasons, we deal with the case where μ is bounded on $\mathscr{J}$.

LEMMA 2.12.– Let X be a submartingale and $J \in \mathscr{J}$. Let $T := D_J$ be the début of the set J, and take $u \in \mathbb{R}_+$ such that $D \subseteq \Omega \times [0, u]$. Then $\mu(J) \leqslant \mathsf{E}(X_u - X_{T \wedge u})$. If random variables X_t converge in L^1 to a random variable X_∞, then $\mu(J) \leqslant \mathsf{E}(X_\infty - X_T)$.

PROOF.– Since T takes a finite number of values, $]\!]T \wedge u, u]\!] \in \mathscr{J}$ and $X(]\!]T \wedge u, u]\!]) = X_u - X_{T \wedge u}$. However, $\mu(]\!]0]\!]) = 0$ and $J \setminus]\!]0]\!] \subseteq]\!]T \wedge u, u]\!]$. The first assertion follows now because μ is nonnegative. The second assertion follows from

$$(X_\infty - X_T) - (X_u - X_{T \wedge u}) = (X_\infty - X_u)\mathbb{1}_{\{T \leqslant u\}}. \qquad \square$$

The most difficult part of the proposed proof of the existence of the Doob–Meyer decomposition is the following lemma.

LEMMA 2.13.– Let X be a submartingale from the class (D). The function $\mu\colon \mathscr{J} \to \mathbb{R}_+$ introduced above is bounded from above and continuous at $\varnothing$, i.e. $J_n \in \mathscr{J}$, $J_1 \supseteq J_2 \supseteq \cdots \supseteq J_n \supseteq \ldots, \cap_n J_n = \varnothing$ imply $\lim_n \mu(J_n) = 0$.

PROOF.– We begin with some preliminary observations. The supermartingale $-X$ satisfies the assumptions of theorem 2.2. Hence, a.s. there exists a limit $\lim_{t \to \infty} X_t =: X_\infty$. Due to the uniform integrability of X, convergence of X_t to X_∞ holds in L^1 as well. Passing to a limit, as $s \to \infty$, in the inequality

$$\int_B X_t \, d\mathsf{P} \leqslant \int_B X_s \, d\mathsf{P}, \quad B \in \mathscr{F}_t, \quad t < s,$$

we get $X_t \leqslant \mathsf{E}(X_\infty | \mathscr{F}_t)$. Hence, $-X$ satisfies assumptions of theorem 2.3.

In particular, $\mathsf{E}X_T \geqslant \mathsf{E}X_0$ for every stopping time T. It follows from lemma 2.12 that μ is bounded from above.

Next, since $X \in (D)$, random variables $X_{T \wedge n}$ converge to X_T in L^1 as $n \to \infty$, where T is an arbitrary stopping time. By theorem A.5 (1), the family X_T, where T runs over the class of all stopping times, is uniformly integrable.

Let (J_n) be a decreasing sequence of sets from $\mathscr{J}$ with empty intersection. Without loss of generality, we assume that $J_n \cap [\![0]\!] = \varnothing$. Fix an $\varepsilon > 0$. Since X is right-continuous in L^1, there are sets $K_n \in \mathscr{J}$ such that the closure $\overline{K_n}$ (the set whose ω-section is the closure in t of the ω-section of K_n for every ω) is in J_n and

$$\mu(J_n) \leqslant \mu(K_n) + 2^{-n}\varepsilon.$$

Put $L_n := K_1 \cap \cdots \cap K_n$; then, for every n,

$$\mu(J_n) \leqslant \mu(L_n) + \varepsilon.$$

The sets $\overline{L_n}$ are decreasing and have empty intersection. Therefore, if T_n is the début of the set $\overline{L_n}$ (that coincides with the début of the set L_n), then the sequence $(T_n(\omega))$ for every ω is monotone and tends to $+\infty$ (and even equals $+\infty$ for large n), which follows from the closedness and the uniform boundedness in t of ω-sections of $\overline{L_n}(\omega)$. By lemma 2.12,

$$\mu(L_n) \leqslant \mathsf{E}(X_\infty - X_{T_n}).$$

Since X belongs to the class (D), the expression on the right tends to 0 as $n \to \infty$ due to the above remark. This yields $\lim_n \mu(J_n) \leqslant \varepsilon$. Since $\varepsilon > 0$ is arbitrary, $\lim_n \mu(J_n) = 0$. $\qquad\square$

THEOREM 2.23.– Let X be a submartingale from the class (D). Then there exists a unique (up to an evanescent set) decomposition $X = M + A$, where $M \in \mathscr{M}$, $A \in \mathscr{A}^+$, and A is predictable.

PROOF.– It is well known from the measure theory that lemma 2.13 allows us to extend the function μ from the ring $\mathscr{J}$ to a finite measure on the σ-algebra $\mathscr{P}$ generated by it, in a unique way. This extension is denoted by the same letter μ.

Let B be an evanescent predictable set. Then its début D_B a.s. equals $+\infty$, hence $C := \{D_B < +\infty\} \in \mathscr{F}_0$. Since $B \subseteq [\![0]\!] \cup]\!]0_C, +\infty[\![,]\!]0_C, +\infty[\![= \cup_n]\!]0_C, n]\!]$, $]\!]0_C, n]\!] = C \times]0, n] \in \mathscr{R}$ and $\mu(]\!]0_C, n]\!]) = \mathsf{E}\mathbb{1}_C(X_n - X_0) = 0$, we have $\mu(B) = 0$.

Thus, the measure μ on $\mathscr{P}$ satisfies the assumptions of theorem 2.17, which implies the existence of a predictable process $A \in \mathscr{A}^+$ such that μ and the restriction of the Doléans measure μ_A onto $\mathscr{P}$ coincide. In particular, put $M := X - A$. Let $s < t$, $B \in \mathscr{F}_s$. Then

$$\mathsf{E}\mathbb{1}_B(M_t - M_s) = \mathsf{E}\mathbb{1}_B(X_t - X_s) - \mathsf{E}\mathbb{1}_B(A_t - A_s)$$
$$= \mu(B \times]s, t]) - \mu_A(B \times]s, t]) = 0.$$

Hence, M is a martingale. Moreover, all A_t are majorized by an integrable random variable A_∞. Therefore, $A \in (D)$ and, hence, $M \in (D)$, see theorem A.5 (2).

Finally, the uniqueness of the decomposition was mentioned above. $\square$

THEOREM 2.24.– Let X be a submartingale. A decomposition $X = M + A$, where $M \in \overline{\mathscr{M}}$, $A \in \mathscr{V}^+$, exists if and only if $X \in (DL)$. In this case, we can choose a predictable A, and such a decomposition is unique (up to an evanescent set).

It turns out that an arbitrary submartingale admits a decomposition if M is allowed to be a local martingale. Such a decomposition is also called the Doob–Meyer decomposition.

THEOREM 2.25.– Let X be a submartingale. Then there exists a unique (up to indistinguishability) decomposition $X = M + A$, where $M \in \mathscr{M}_{\mathrm{loc}}$, $A \in \mathscr{V}^+$, and A is predictable.

PROOF OF THEOREM 2.25.– Define a stopping time T_n by $T_n := \inf\{t: |X_t| > n\} \wedge n$. Note that T_n increase to $+\infty$. Then $X^*_{T_n} \leqslant n + |X_{T_n}|$, and X_{T_n} is integrable by corollary 2.1. Hence, $X^{T_n} \in (D)$. However, X^{T_n} is a submartingale by corollary 2.2. By theorem 2.23, for every n, there is a decomposition

$$X^{T_n} = M^n + A^n, \tag{2.37}$$

where $M^n \in \mathscr{M}$, $A^n \in \mathscr{A}^n$, A^n is predictable for every n. Stopping processes in both sides of [2.37] at time T_n and comparing with [2.37], we get from the uniqueness of the Doob–Meyer decomposition that

$$A^n \text{ and } (A^n)^{T_n} \text{ are indistinguishable for every } n = 1, 2, \dots. \tag{2.38}$$

Similarly, stopping processes in both sides of [2.37] at time T_{n-1} and comparing with [2.37] for index $n - 1$, we get from the uniqueness of the Doob–Meyer decomposition that

$$A^{n-1} \text{ and } (A^n)^{T_{n-1}} \text{ are indistinguishable for every } n = 2, 3, \dots. \tag{2.39}$$

Now let A be obtained by the "gluing" procedure from $\{A^n\}$, $\{T_n\}$. Due to proposition 2.9, A is a predictable increasing process. Moreover,

A^n and A^{T_n} are indistinguishable for every $n = 1, 2, \ldots$.

Therefore, if $M := X - A$, then

M^n and M^{T_n} are indistinguishable for every $n = 1, 2, \ldots$.

Hence, $M \in \mathcal{M}_{\mathrm{loc}}$.

The uniqueness of the decomposition was mentioned at the beginning of this section. $\square$

PROOF OF THEOREM 2.24.– The existence of the decomposition is proved similar to that in theorem 2.25 with the only difference that one should take $T_n := n$. The necessity of the condition and the uniqueness of the decomposition were discussed earlier. $\square$

EXERCISE 2.40.– Let X be a submartingale and a process with independent increments on the considered stochastic basis. Find its Doob–Meyer decomposition.

HINT.– Use exercise 2.6.

2.7. Square-integrable martingales

Unless otherwise stated, we will assume that a stochastic basis $\mathbb{B} = (\Omega, \mathscr{F}, \mathbb{F} = (\mathscr{F}_t)_{t \in \mathbb{R}_+}, \mathsf{P})$ satisfying the usual conditions is given.

DEFINITION 2.17.– A martingale from the space $\mathcal{M}^2$ is called a *square-integrable martingale*.

Recall that

$$\mathcal{M}^2 = \left\{ M \in \mathcal{M} : \mathsf{E}M_\infty^2 < \infty \right\} = \left\{ M \in \overline{\mathcal{M}} : \sup_{t \in \mathbb{R}_+} \mathsf{E}M_t^2 < \infty \right\}$$

$$= \left\{ M \in \mathcal{M}_{\mathrm{loc}} : \mathsf{E}(M_\infty^*)^2 < \infty \right\},$$

see section 2.1 and theorem 2.8. In particular,

$$\mathsf{E}(MN)_\infty^* < +\infty \quad \text{for every } M, N \in \mathcal{M}^2. \tag{2.40}$$

We continue to speak about elements of the space $\mathcal{M}^2$ as stochastic processes. But the reader should keep in mind that in this section, as a rule, elements of $\mathcal{M}^2$ are to be interpreted as equivalence classes consisting of indistinguishable square-integrable martingales. As already noted, then $\mathscr{M}^2$ is isomorphic to $L^2(\mathscr{F}_\infty)$ and, therefore, is a Hilbert space with the scalar product

$$(M, N)_{\mathscr{M}^2} := \mathsf{E} M_\infty N_\infty.$$

Orthogonality in this sense will be called weak orthogonality in order to emphasize its difference from the notion of strong orthogonality introduced below.

Recall also that

$$\|M\|_{\mathscr{M}^2} \leqslant \|M\|_{\mathscr{H}^2} \leqslant 2\|M\|_{\mathscr{M}^2}, \quad M \in \mathscr{M}^2, \qquad [2.41]$$

where

$$\|M\|_{\mathscr{H}^2}^2 = \mathsf{E}(M_\infty^*)^2.$$

If a sequence converges in L^2, it converges in probability, hence, there is a subsequence converging almost surely. Using this argument, we obtain, in particular, the following lemma which will often be used in this section.

LEMMA 2.14.– If a sequence (M^n) in $\mathscr{M}^2$ converges to M in the norm of $\mathscr{M}^2$, then there is a subsequence (n_k) such that

$$\lim_{k \to \infty} (M - M^{n_k})_\infty^* = 0 \quad \text{a.s.,}$$

i.e. for almost all ω, trajectories $M_\cdot^{n_k}(\omega)$ converge to $M_\cdot(\omega)$ uniformly in t.

EXERCISE 2.41.– Let $M \in \mathscr{M}^2$. Show that, for $s < t$,

$$\mathsf{E}(M_t^2 - M_s^2|\mathscr{F}_s) = \mathsf{E}((M_t - M_s)^2|\mathscr{F}_s).$$

Let $M \in \mathscr{M}^2$. Consider the square M^2 of M. By proposition 2.1, see also the previous exercise, M^2 is a submartingale. Moreover, $M^2 \in (D)$ by [2.40]. Hence, we can apply theorem 2.23, which asserts that there exists the Doob–Meyer decomposition of M^2:

$$M^2 = N + A, \quad N \in \mathscr{M}, \quad A \in \mathscr{A}^+, \quad A \text{ is predictable.}$$

DEFINITION 2.18.– *The quadratic characteristic of a square-integrable martingale* $M \in \mathscr{M}^2$ *is a predictable integrable increasing process, denoted by* $\langle M, M \rangle$ *or* $\langle M \rangle$, *such that* $M^2 - \langle M, M \rangle$ *is a uniformly integrable martingale. The mutual quadratic characteristic of square-integrable martingales* $M, N \in \mathscr{M}^2$ *is a predictable process with integrable variation, denoted by* $\langle M, N \rangle$, *such that* $MN - \langle M, N \rangle$ *is a uniformly integrable martingale.*

It is clear that the quadratic characteristic $\langle M, M \rangle$ and the mutual quadratic characteristic $\langle M, N \rangle$ do not depend on the choice of versions of M and N. It follows from the argument given before the definition that the quadratic characteristic $\langle M, M \rangle$ exists for every $M \in \mathcal{M}^2$, while corollary 2.14 implies its uniqueness (up to indistinguishability). The mutual quadratic characteristic $\langle M, N \rangle$ is also unique (up to indistinguishability) by the same corollary, which yields the following fact as well: if A is a predictable process, $A \in \mathcal{V}$ (respectively $A \in \mathcal{V}^+$) and $MN - A \in \mathcal{M}_{\mathrm{loc}}$ (respectively $M^2 - A \in \mathcal{M}_{\mathrm{loc}}$), then $A = \langle M, N \rangle$ (respectively $A = \langle M, M \rangle$). The existence of the mutual quadratic characteristic $\langle M, N \rangle$ for every $M, N \in \mathcal{M}^2$ is proved using polarization: for example, we can take the process

$$\frac{1}{2}(\langle M + N, M + N \rangle - \langle M, M \rangle - \langle N, N \rangle)$$

as $\langle M, N \rangle$.

The quadratic characteristic $\langle M, M \rangle$ is often called the *angle bracket* of a martingale $M \in \mathcal{M}^2$.

Since the quadratic characteristic is determined up to indistinguishability, relations between them will always be understood up to an evanescent set, and it will not always be mentioned explicitly.

Let us note that, obviously, the form $\langle M, N \rangle$ is symmetric and bilinear in the sense just indicated.

EXERCISE 2.42.– Let a martingale $M \in \mathcal{M}^2$ be a process with independent increments on the considered basis. Find $\langle M, M \rangle$.

LEMMA 2.15.– Let $M, N \in \mathcal{M}^2$ and T be a stopping time. Then $\langle M, N^T \rangle = \langle M, N \rangle^T$.

PROOF.– Put $L := MN^T - (MN)^T$. We will check that $L \in \mathcal{M}$ using lemma 2.10. It is enough to verify only the last assumption on L. Let S be a stopping time. Then

$$\mathsf{E} L_S = \mathsf{E}(M_S N_{S \wedge T} - M_{S \wedge T} N_{S \wedge T})$$
$$= \mathsf{E}(\mathsf{E}(M_S | \mathscr{F}_{S \wedge T}) N_{S \wedge T} - M_{S \wedge T} N_{S \wedge T}) = 0,$$

because $\mathsf{E}(M_S | \mathscr{F}_{S \wedge T}) = M_{S \wedge T}$ by theorem 2.4.

Thus, $L \in \mathcal{M}$. On the other hand, $(MN)^T - \langle M, N \rangle^T$ is a uniformly integrable martingale by the definition of the characteristic. Therefore, $MN^T - \langle M, N \rangle^T \in \mathcal{M}$ □

THEOREM 2.26.– Let $M, N \in \mathcal{M}^2$, M and N be a.s. continuous. Then $\langle M, N \rangle$ has a continuous version.

PROOF.– This proof follows from theorems 1.18 and 2.4, see the proof of theorem 2.18. $\square$

DEFINITION 2.19.– Square-integrable martingales $M, N \in \mathcal{M}^2$ are called *strongly orthogonal*, if $M_0 N_0 = 0$ a.s. and MN is a local martingale.

LEMMA 2.16.–

1) If $M, N \in \mathcal{M}^2$ are strongly orthogonal, then $MN \in \mathcal{M}$.

2) If $M, N \in \mathcal{M}^2$ are strongly orthogonal, then they are weakly orthogonal.

3) $M, N \in \mathcal{M}^2$ are strongly orthogonal if and only if $M_0 N_0 = 0$ a.s. and $\langle M, N \rangle = 0$.

PROOF.– (1) This proof follows from [2.40] and theorem 2.8. To prove (2) it is enough to note that $\mathsf{E} M_\infty N_\infty = \mathsf{E} M_0 N_0 = 0$ by part (1). Finally, (3) follows directly from the definitions. $\square$

DEFINITION 2.20.– A linear subspace $\mathcal{H} \subseteq \mathcal{M}^2$ is called a *stable subspace* if

1) $\mathcal{H}$ is closed in the norm $\| \cdot \|_{\mathcal{M}^2}$;

2) $\mathcal{H}$ is stable under stopping, i.e. $M \in \mathcal{H}$ implies $M^T \in \mathcal{H}$ for every stopping time T;

3) $M \in \mathcal{H}, B \in \mathcal{F}_0$ implies $M \mathbb{1}_B \in \mathcal{H}$.

Recall that, if L is a closed linear subspace in a Hilbert space H, the set $L^\perp$ of vectors that are orthogonal to all vectors in L, is called the orthogonal complement of L. It is known that $L^\perp$ is a closed linear subspace itself, and any vector in H has a unique decomposition into the sum of a vector from L and a vector from $L^\perp$.

THEOREM 2.27.– Let $\mathcal{H} \subseteq \mathcal{M}^2$ be a stable subspace, and let $\mathcal{H}^\perp := \{N \in \mathcal{M}^2 : \mathsf{E} M_\infty N_\infty = 0 \text{ for every } M \in \mathcal{H}\}$ be its orthogonal complement. Then $\mathcal{H}^\perp$ is a stable subspace, and M and N are strongly orthogonal for every $M \in \mathcal{H}, N \in \mathcal{H}^\perp$.

COROLLARY 2.15.– Let $\mathcal{H} \subseteq \mathcal{M}^2$ be a stable subspace. Then every $M \in \mathcal{M}^2$ has a unique decomposition $M = M' + M''$, where $M' \in \mathcal{H}$ and M'' is strongly orthogonal to all martingales in $\mathcal{H}$.

PROOF OF THEOREM 2.27.– As mentioned before the statement, $\mathcal{H}^\perp$ is closed in the norm according to the theory of Hilbert spaces.

Let us take an arbitrary $N \in \mathcal{H}^\perp$. Then

$$\mathsf{E} M_\infty N_\infty = 0 \quad \text{for every } M \in \mathcal{H}. \tag{2.42}$$

Let $M \in \mathcal{H}$. For any stopping time T, we have $M^T \in \mathcal{H}$ and we can apply [2.42] with M^T instead of M:

$$\mathsf{E} M_T N_T = \mathsf{E}\big(M_T \mathsf{E}(N_\infty | \mathscr{F}_T)\big) = \mathsf{E} M_T N_\infty = 0. \qquad [2.43]$$

Therefore, $MN \in \mathcal{M}$ by lemma 2.10. Moreover, it follows from [2.43] that

$$\mathsf{E} M_\infty N_T = \mathsf{E}\big(\mathsf{E}(M_\infty N_T | \mathscr{F}_T)\big) = \mathsf{E} M_T N_T = 0.$$

Hence, $N^T \in \mathcal{H}^\perp$ because $M \in \mathcal{H}$ is arbitrary.

Let $M \in \mathcal{H}$ and $B \in \mathscr{F}_0$. Since $M\mathbb{1}_B \in \mathcal{H}$, we can apply [2.42] with $M\mathbb{1}_B$ instead of M:

$$\mathsf{E}\mathbb{1}_B M_\infty N_\infty = 0. \qquad [2.44]$$

Hence, $N\mathbb{1}_B \in \mathcal{H}^\perp$ because $M \in \mathcal{H}$ is arbitrary. However, we have already proved that $\mathsf{E}(M_\infty N_\infty | \mathscr{F}_0) = M_0 N_0$. Therefore, it follows from [2.44] that

$$\mathsf{E}\mathbb{1}_B M_0 N_0 = \mathsf{E}\mathbb{1}_B \mathsf{E}(M_\infty N_\infty | \mathscr{F}_0) = 0,$$

and $M_0 N_0 = 0$ a.s. because $B \in \mathscr{F}_0$ is arbitrary. $\qquad \square$

Here are some examples of stable subspaces.

EXAMPLE 2.5.– Let us define the subspace $\mathcal{M}^{2,c}$ of continuous square-integrable martingales as the set of all elements (equivalence classes) in $\mathcal{M}^2$ which contain a continuous square-integrable martingale, i.e. consist of a.s. continuous square-integrable martingales. Less formally,

$$\mathcal{M}^{2,c} := \{M \in \mathcal{M}^2 : \text{trajectories } M_.(\omega) \text{ are a.s. continuous.}\}$$

$\mathcal{M}^{2,c}$ is a stable subspace. Indeed, that this space is linear and properties (2), (3) in the definition of a stable subspace are obvious. Property (1) follows from lemma 2.14. The orthogonal complement of $\mathcal{M}^{2,c}$ is denoted by $\mathcal{M}^{2,d}$ and is called the subspace of *purely discontinuous* square-integrable martingales.

Let us note that the process identically equal to 1 is in $\mathcal{M}^{2,c}$. Hence, since elements of $\mathcal{M}^{2,c}$ and $\mathcal{M}^{2,d}$ are strongly orthogonal, we have $N_0 = 0$ a.s. for every $N \in \mathcal{M}^{2,d}$.

By corollary 2.15, every martingale $M \in \mathcal{M}^2$ has a unique (up to indistinguishability) decomposition

$$M = M_0 + M^c + M^d, \quad M^c \in \mathcal{M}^{2,c}, \quad M^d \in \mathcal{M}^{2,d}. \qquad [2.45]$$

The components M^c and M^d of this decomposition are called the *continuous martingale component* and the *purely discontinuous martingale component* of M. It follows from the definition that $M_0^c = M_0^d = 0$. The uniqueness of this decomposition implies that, for every stopping time T, up to indistinguishability,

$$(M^T)^c = (M^c)^T, \quad (M^T)^d = (M^d)^T.$$

The reader should be warned that, for a square-integrable martingale with finite variation, decomposition [2.45] into continuous and purely discontinuous *martingale* components, as a rule, differs from the decomposition into continuous and purely discontinuous processes with finite variation in exercise 2.22. See, in particular, theorem 2.28 below.

EXAMPLE 2.6.– Let T be a strictly positive stopping time. Put

$$\mathscr{M}^2[T] := \{ M \in \mathscr{M}^{2,d} : \{\Delta M \neq 0\} \setminus [\![T]\!] \text{ is an evanescent set}\}.$$

$\mathscr{M}^2[T]$ is a stable subspace: again, the linearity and properties (2), (3) from the definition of stable subspaces are obvious and property (1) follows from lemma 2.14.

Note that the intersection of any family of stable subspaces is a stable subspace. Consequently, there exists the smallest stable subspace containing a given martingale $M \in \mathscr{M}^2$, and it will be described in Chapter 3. A somewhat more difficult problem is a description of the smallest stable subspace containing a finite set of martingales $M^1, \ldots, M^k \in \mathscr{M}^2$. Let us only mention here that it does not coincide, in general, with the sum of smallest stable subspace containing M^j, $j = 1, \ldots, k$, which may not be closed in the norm.

THEOREM 2.28.– $\mathscr{A} \cap \mathscr{M}^2 \subseteq \mathscr{M}^{2,d}$.

PROOF.– Let $M \in \mathscr{A} \cap \mathscr{M}^2$. By lemma 2.11, M is orthogonal to any bounded continuous martingale. Since every continuous $N \in \mathscr{M}^{2,c}$ is a limit (in $\mathscr{M}^2$) of bounded martingales N^{T_n}, where $T_n := \inf \{t : |N_t| > n\}$, the claim follows.   □

The following example shows that the inclusion in theorem 2.28 is strict in general.

EXAMPLE 2.7.– Let $\xi_1, \ldots, \xi_n, \ldots$ be independent random variables on a complete probability space $(\Omega, \mathscr{F}, \mathrm{P})$ and $\mathrm{P}(\xi_n = \pm 1/n) = 1/2$. Put $\mathscr{F}_t := \sigma\{\xi_1, \ldots, \xi_{[t]}, \mathscr{N}\}$, where $[\cdot]$ is the integer part of a number and $\mathscr{N}$ is a family of null sets in $\mathscr{F}$,

$$M_t := \sum_{n \leqslant t} \xi_n.$$

It is clear that $M = (M_t)$ is a martingale and

$$\mathsf{E}M_t^2 = \mathsf{E}\Big(\sum_{n \leqslant t} \xi_n\Big)^2 = \sum_{n \leqslant t} \mathsf{E}\xi_n^2 = \sum_{n \leqslant t} \frac{1}{n^2} \leqslant \sum_{n=1}^{\infty} \frac{1}{n^2} < \infty,$$

hence, $M \in \mathscr{M}^2$. Let M^n be a process M stopped at time n. Then, by theorem 2.28, $M^n \in \mathscr{M}^{2,d}$, hence, M, being the limit of M^n in $\mathscr{M}^2$, is also an element on the subspace $\mathscr{M}^{2,d}$. However,

$$\mathrm{Var}\,(M)_\infty = \sum_{n=1}^{\infty} |\xi_n| = \sum_{n=1}^{\infty} \frac{1}{n} = \infty.$$

We need the following two lemmas in order to study purely discontinuous square-integrable martingales.

LEMMA 2.17.– Let T be a totally inaccessible strictly positive stopping time. Then $\mathscr{M}^2[T] \subseteq \mathscr{M}_{\mathrm{loc}} \cap \mathscr{V}$ and the subspace $\mathscr{M}^2[T]$ consists precisely of processes of the form $M = A - \widetilde{A}$, where $A = \xi \mathbb{1}_{[T,\infty[}$, ξ is a square-integrable $\mathscr{F}_T$-measurable random variable vanishing on $\{T = \infty\}$, $\widetilde{A}$ is the compensator of A, and $\widetilde{A}$ is continuous. If $M \in \mathscr{M}^2[T]$ and $N \in \mathscr{M}^2$, then

$$MN - \Delta M_T \Delta N_T \mathbb{1}_{[T,\infty[} \in \mathscr{M}.$$

In particular, $\mathsf{E}M_\infty^2 = \mathsf{E}(\Delta M_T)^2 \mathbb{1}_{\{T<\infty\}}$. If $N \in \mathscr{M}^2$, then the projection M of a martingale N onto $\mathscr{M}^2[T]$ has the indicated form with $\xi = \Delta N_T \mathbb{1}_{\{T<\infty\}}$.

PROOF.– Let M have the form indicated in the lemma, then $M \in \mathscr{M}_{\mathrm{loc}}$. In order to prove that $M \in \mathscr{M}^2$, it is enough to check that $\mathsf{E}(M_\infty^*)^2 < \infty$. To do this, in its turn, it is enough to consider nonnegative ξ and to show that $\mathsf{E}\widetilde{A}_\infty^2 < \infty$.

Let η be any $\mathscr{F}_\infty$-measurable bounded nonnegative random variable and L a bounded martingale with $L_\infty = \eta$. By lemma 2.3, proposition 2.7, the Schwarz inequality and Doob's inequality (corollary 2.6),

$$\mathsf{E}L_\infty \widetilde{A}_\infty = \mathsf{E}(L_- \cdot \widetilde{A}_\infty) = \mathsf{E}(L_- \cdot A_\infty) \leqslant \mathsf{E}L_\infty^* A_\infty$$

$$\leqslant \Big(\mathsf{E}\big(L_\infty^*\big)^2 \mathsf{E}A_\infty^2\Big)^{1/2} \leqslant 2\Big(\mathsf{E}L_\infty^2 \mathsf{E}A_\infty^2\Big)^{1/2}.$$

Since

$$\big(\mathsf{E}\widetilde{A}_\infty^2\big)^{1/2} = \sup_{\substack{\eta \in L^2(\mathscr{F}_\infty),\, \eta \geqslant 0 \\ \mathsf{E}\eta^2 = 1}} \mathsf{E}\eta \widetilde{A}_\infty = \sup_{\substack{\eta \in L^2(\mathscr{F}_\infty),\, \eta \geqslant 0 \\ \eta \text{ bounded},\, \mathsf{E}\eta^2 = 1}} \mathsf{E}\eta \widetilde{A}_\infty,$$

we get

$$\mathsf{E}\widetilde{A}_\infty^2 \leqslant 4\mathsf{E}A_\infty^2 = 4\mathsf{E}\xi^2 < \infty.$$

Thus, $M \in \mathscr{M}^2$. By lemma 2.8, the compensator $\widetilde{A}$ is a.s. continuous. Hence, the set $\{\Delta M \neq 0\} \setminus [\![T]\!]$ is evanescent. By lemma 2.11,

$$\mathsf{E}M_\infty N_\infty = \mathsf{E}\xi\Delta N_T \mathbb{1}_{\{T<\infty\}} \qquad\qquad [2.46]$$

for every bounded martingale N. Next, bounded martingales are dense in $\mathscr{M}^2$ and both sides of [2.46] are continuous linear functionals in N on $\mathscr{M}^2$ (the right side is continuous in N because the mapping $N \rightsquigarrow \Delta N_T \mathbb{1}_{\{T<\infty\}}$ from $\mathscr{M}^2$ to L^2 is continuous due to the inequalities $(\Delta N)^* \leqslant 2N^*$ and [2.41]). Hence, [2.46] is valid for every $N \in \mathscr{M}^2$. In particular, M is weakly orthogonal to continuous N, i.e. $M \in \mathscr{M}^{2,d}$, hence $M \in \mathscr{M}^2[T]$.

Take any $N \in \mathscr{M}^2[T]$ and put $\xi := \Delta N_T \mathbb{1}_{\{T<\infty\}}$. Define M as in the statement of the lemma. Then, on the one hand, $N - M$ is a.s. continuous, i.e. $N - M \in \mathscr{M}^{2,c}$; on the other hand, $N - M \in \mathscr{M}^2[T] \subseteq \mathscr{M}^{2,d}$. Therefore, $N = M$, i.e. the subspace $\mathscr{M}^2[T]$ consists only of martingales M of the considered form. In particular, $\mathscr{M}^2[T] \subseteq \mathscr{V}$.

Take any $N \in \mathscr{M}^2$ and put $\xi := \Delta N_T \mathbb{1}_{\{T<\infty\}}$. Again, define M as in the previous paragraph. Then $\Delta(N - M)_T \mathbb{1}_{\{T<\infty\}} = 0$ a.s. Now it follows from [2.46] that $N - M$ is weakly orthogonal to any element in $\mathscr{M}^2[T]$. Hence, M is the projection of N onto $\mathscr{M}^2[T]$.

Let $M \in \mathscr{M}^2[T]$ and $N \in \mathscr{M}^2$. Put $L := MN - \Delta M_T \Delta N_T \mathbb{1}_{[\![T,\infty[\![}$ and let S be an arbitrary stopping time. Applying [2.46] with N^S instead of N, we get

$$\mathsf{E}M_S N_S = \mathsf{E}M_\infty N_S = \mathsf{E}\Delta M_T \Delta N_T \mathbb{1}_{\{S\geqslant T,\, T<\infty\}},$$

i.e. $\mathsf{E}L_S = 0$. By lemma 2.10, $L \in \mathscr{M}$. □

LEMMA 2.18.– Let T be a predictable strictly positive stopping time. Then $\mathscr{M}^2[T] \subseteq \mathscr{M}_{\mathrm{loc}} \cap \mathscr{V}$ and the subspace $\mathscr{M}^2[T]$ consists precisely of processes of the form $M = \xi \mathbb{1}_{[\![T,\infty[\![}$, where ξ is a square-integrable $\mathscr{F}_T$-measurable random variable, vanishing on $\{T = \infty\}$ and such that $\mathsf{E}(\xi|\mathscr{F}_{T-}) = 0$. If $M \in \mathscr{M}^2[T]$ and $N \in \mathscr{M}^2$, then

$$MN - \Delta M_T \Delta N_T \mathbb{1}_{[\![T,\infty[\![} \in \mathscr{M}.$$

In particular, $\mathsf{E}M_\infty^2 = \mathsf{E}(\Delta M_T)^2 \mathbb{1}_{\{T<\infty\}}$. If $N \in \mathscr{M}^2$, then the projection M of a martingale N onto $\mathscr{M}^2[T]$ has the indicated form with $\xi = \Delta N_T \mathbb{1}_{\{T<\infty\}}$.

PROOF.– If M has the form indicated in the lemma, then $M \in \mathscr{M}_{\mathrm{loc}}$ by lemma 2.9; therefore, $M \in \mathscr{M}^2$. From now on, repeat the proof of lemma 2.17.   □

COROLLARY 2.16.– Let T be a predictable or totally inaccessible strictly positive stopping time. Then a square-integrable martingale N belongs to the orthogonal complement of $\mathscr{M}^2[T]$ if and only if $\Delta N_T \mathbb{1}_{\{T<\infty\}} = 0$ a.s.

In the next theorem, we describe a structure of the purely discontinuous martingale component of any square-integrable martingale.

THEOREM 2.29.– Let $M \in \mathscr{M}^2$ and let (T_n) be a sequence of stopping times such that every T_n is strictly positive and is either predictable or totally inaccessible,

$$\{\Delta M \neq 0\} \subseteq \bigcup_n [\![T_n]\!] \qquad [2.47]$$

and

$$[\![T_n]\!] \cap [\![T_m]\!] = \varnothing, \quad m \neq n. \qquad [2.48]$$

Put $A^n = \Delta M_{T_n} \mathbb{1}_{[\![T_n,\infty[\![}}$, $M^n := A^n - \widetilde{A}^n$, $N^n := M^1 + \cdots + M^n$, where $\widetilde{A}^n$ is the compensator of A^n. Then the sequence N^n converges in $\mathscr{M}^2$ to M^d as $n \to \infty$. Moreover,

$$\mathsf{E}(M_\infty^d)^2 = \mathsf{E} \sum_{s \in \mathbb{R}_+} (\Delta M_s)^2. \qquad [2.49]$$

REMARK 2.11.– A sequence (T_n) of stopping times satisfying the assumptions of the theorem always exists; see theorem 1.17.

PROOF.– We have $M^n \in \mathscr{M}^2[T_n]$ by lemmas 2.17 and 2.18, $\Delta M^n = \Delta A^n = \Delta M \mathbb{1}_{[\![T_n]\!]}$, and the graphs $[\![T_n]\!]$ are disjoint. Hence, the martingales $M^1, \ldots, M^n, \ldots$ are strongly orthogonal by corollary 2.16. By the same reason, $M - N^n$ is strongly orthogonal to $M^1, \ldots, M^n$ and, hence, to their sum N^n. Therefore,

$$\mathsf{E}M_\infty^2 = \mathsf{E}(N_\infty^n)^2 + \mathsf{E}(M_\infty - N_\infty^n)^2 = \sum_{k=1}^n \mathsf{E}(M_\infty^k)^2 + \mathsf{E}(M_\infty - N_\infty^n)^2$$

$$= \sum_{k=1}^n \mathsf{E}(\Delta M_{T_k})^2 \mathbb{1}_{\{T_k<\infty\}} + \mathsf{E}(M_\infty - N_\infty^n)^2, \qquad [2.50]$$

where we use lemmas 2.17 and 2.18 in the last equality. In particular, $\sum_{k=1}^\infty \mathsf{E}(M_\infty^k)^2 < \infty$. By the Cauchy criterion, the orthogonal series $\sum_{k=1}^\infty M^k$

converges in $\mathcal{M}^2$. Denote its sum by N. Since N is the limit of purely discontinuous square-integrable martingales N^n, we have $N \in \mathcal{M}^{2,d}$. On the other hand, $\Delta N^n = \Delta M \mathbb{1}_{\cup_{k=1}^n [\![T_k]\!]}$. Therefore, it follows from [2.47] and lemma 2.14 that ΔM and ΔN are indistinguishable, i.e. $M - N \in \mathcal{M}^{2,c}$. Thus, $N = M^d$ and $M - N = M_0 + M^c$.

Finally, passing to the limit as $n \to \infty$ in [2.50] and using [2.47] and [2.48], we get

$$\mathsf{E}(M_\infty^d)^2 + \mathsf{E}(M_0 + M_\infty^c)^2 = \mathsf{E}M_\infty^2$$
$$= \sum_{k=1}^\infty \mathsf{E}(\Delta M_{T_k})^2 \mathbb{1}_{\{T_k < \infty\}} + \mathsf{E}(M_\infty - N_\infty)^2$$
$$= \mathsf{E} \sum_{s \in \mathbb{R}_+} (\Delta M_s)^2 + \mathsf{E}(M_0 + M_\infty^c)^2,$$

hence [2.49] follows. $\qquad\square$

COROLLARY 2.17.– A purely discontinuous square-integrable martingale M is orthogonal to a square-integrable martingale N if $\Delta M \Delta N = 0$ (up to an evanescent set).

PROOF.– Take the sequence (M^n) as in theorem 2.29, then M^n is orthogonal to N for every n by corollary 2.16, while M is the limit of the sums $M^1 + \cdots + M^n$ by theorem 2.29. $\qquad\square$

According to [2.49],

$$\mathsf{E} \sum_{s \in \mathbb{R}_+} (\Delta M_s)^2 < +\infty$$

for every $M \in \mathcal{M}^2$. However, formally, the process $\sum_{s \leqslant .} (\Delta M_s)^2$, in general, may not be in $\mathcal{A}^+$ and may not be increasing, because it may take value $+\infty$ and not be right-continuous at one point (on a null set). We will often encounter such processes in the rest of the book. So let us introduce a formal definition.

DEFINITION 2.21.– Let X be an optional process such that

$$\{X \neq 0\} \subseteq \bigcup_n [\![T_n]\!]$$

for a sequence (T_n) of stopping times. Assume that $X_0 = 0$ a.s. and, for every $t \in \mathbb{R}_+$,

$$\sum_{s \leqslant t} |X_s| < +\infty \quad \text{a.s.}$$

Under these assumptions, let us define a process $S(X)$ as a purely discontinuous process with finite variation such that, for every $t \in \mathbb{R}_+$,

$$S(X)_t = \sum_{s \leqslant t} X_s \quad \text{a.s.}$$

LEMMA 2.19.– The process $S(X)$ in definition 2.21 is well defined and unique up to indistinguishability. If X is predictable, then $S(X)$ is predictable. If $X \geqslant 0$, then $S(X)$ has a version in $\mathscr{V}^+$.

PROOF.– Without loss of generality, we may assume that the graphs $[\![T_n]\!]$ are disjoint. Then, for every $t \in \mathbb{R}_+$,

$$\sum_{s \leqslant t} |X_s| = \sum_{n=1}^{\infty} |X_{T_n}| \mathbb{1}_{\{T_n \leqslant t\}}, \quad \sum_{s \leqslant t} X_s = \sum_{n=1}^{\infty} X_{T_n} \mathbb{1}_{\{T_n \leqslant t\}},$$

which implies that

$$\left\{ \sum_{s \leqslant t} |X_s| < +\infty \right\} \in \mathscr{F}_t$$

and the random variable

$$\sum_{s \leqslant t} X_s$$

is $\mathscr{F}_t$-measurable. So put

$$S(X)_t = \sum_{s \leqslant t} X_s$$

on the set $B := \bigcap_n \left\{ \sum_{s \leqslant n} |X_s| < +\infty \right\} \cap \{X_0 = 0\}$ of measure one and $S(X)_t = 0$ on its complement. Under such a definition, $S(X) \in \mathscr{V}^d$ and $\Delta S(X) = X \mathbb{1}_B$. If X is predictable, then this version of $S(X)$ is predictable by proposition 1.8, while an arbitrary version of $S(X)$ is predictable by corollary 1.2. $\square$

DEFINITION 2.22.– The *quadratic variation* of a square-integrable martingale $M \in \mathscr{M}^2$ is the increasing process

$$[M, M] := \langle M^c, M^c \rangle + S\big((\Delta M)^2\big).$$

The *quadratic covariation* of square-integrable martingales $M, N \in \mathscr{M}^2$ is the process with finite variation

$$[M, N] := \langle M^c, N^c \rangle + S(\Delta M \Delta N).$$

The quadratic variation $[M, M]$ is often called the *quadratic bracket* of a martingale $M \in \mathscr{M}^2$.

Since the quadratic covariation is determined up to indistinguishability, all relations including quadratic brackets are always understood up to indistinguishability. In particular, the form $[M, N]$ is symmetric and bilinear in this sense.

Note that the process $\langle M^c, N^c \rangle$ is a.s. continuous by theorem 2.26. Therefore, the processes $\Delta[M, N]$ and $\Delta M \Delta N$ are indistinguishable for every $M, N \in \mathscr{M}^2$.

It follows from the definition and [2.49] that, for $M, N \in \mathscr{M}^2$,

$$[M, M] \in \mathscr{A}^+, \quad [M, N] \in \mathscr{A},$$

and lemma 2.15 implies

$$[M, N^T] = [M, N]^T \qquad\qquad [2.51]$$

for every stopping time T. Furthermore, $(M^c)^2 - \langle M^c, M^c \rangle \in \mathscr{M}$ by the definition of the quadratic characteristic. Taking into account [2.49], we have

$$\mathsf{E} M_\infty^2 = \mathsf{E} M_0^2 + \mathsf{E}(M_\infty^c)^2 + \mathsf{E}(M_\infty^d)^2 = \mathsf{E} M_0^2 + \mathsf{E}[M, M]_\infty.$$

Applying this to M^T and using [2.51], we get $\mathsf{E} M_T^2 = \mathsf{E} M_0^2 + \mathsf{E}[M, M]_T$. By lemma 2.10, we have $M^2 - [M, M] \in \mathscr{M}$. Using polarization, we arrive at the following statement.

LEMMA 2.20.– For $M, N \in \mathscr{M}^2$, the processes $MN - [M, N]$ and $[M, N] - \langle M, N \rangle$ are uniformly integrable martingales. In particular, $\langle M, N \rangle$ is the compensator of $[M, N]$.

Let us formulate without proof a statement which explains the origin of the term "quadratic variation". For every $M \in \mathscr{M}^2$ and $t \in \mathbb{R}_+$ for every sequences of partitions $\gamma_n = \{0 = t_0^n < t_1^n < \cdots < t_{k(n)}^n = t\}$ such that $|\gamma_n| := \max_{k=1,\dots,k(n)} |t_k^n - t_{k-1}^n| \to 0$, it holds

$$[M, M]_t = \mathsf{P}\text{-}\lim_n \sum_{k=1}^{k(n)} \left(M_{t_k^n} - M_{t_{k-1}^n}\right)^2, \qquad\qquad [2.52]$$

where $\mathsf{P}\text{-}\lim$ stands for the limit in probability.

THEOREM 2.30 (Kunita–Watanabe inequalities).– Let $M, N \in \mathscr{M}^2$, and let H and K be measurable processes. Then a.s.

$$\int_0^\infty |H_s K_s|\, d\operatorname{Var} \langle M, N \rangle_s \leqslant \left(\int_0^\infty H_s^2\, d\langle M, M \rangle_s \right)^{1/2} \left(\int_0^\infty K_s^2\, d\langle N, N \rangle_s \right)^{1/2} \quad [2.53]$$

and

$$\int_0^\infty |H_s K_s|\, d\operatorname{Var} [M, N]_s \leqslant \left(\int_0^\infty H_s^2\, d[M, M]_s \right)^{1/2} \left(\int_0^\infty K_s^2\, d[N, N]_s \right)^{1/2}. \quad [2.54]$$

In particular,

$$\mathsf{E} \int_0^\infty |H_s K_s|\, d\operatorname{Var} \langle M, N \rangle_s \leqslant \left(\mathsf{E} \int_0^\infty H_s^2\, d\langle M, M \rangle_s \right)^{1/2}$$

$$\times \left(\mathsf{E} \int_0^\infty K_s^2\, d\langle N, N \rangle_s \right)^{1/2} \quad [2.55]$$

and

$$\mathsf{E} \int_0^\infty |H_s K_s|\, d\operatorname{Var} [M, N]_s \leqslant \left(\mathsf{E} \int_0^\infty H_s^2\, d[M, M]_s \right)^{1/2}$$

$$\times \left(\mathsf{E} \int_0^\infty K_s^2\, d[N, N]_s \right)^{1/2}. \quad [2.56]$$

PROOF.– Let us prove inequality [2.54]. Put

$$A := [M, M] + [N, N] + \operatorname{Var}([M, N]).$$

By theorem 2.20, there exist optional processes F, G and J such that $F, G \geqslant 0$ everywhere and

$$[M, M] = F \cdot A, \quad [N, N] = G \cdot A, \quad [M, N] = J \cdot A$$

up to indistinguishability. Next, let $\lambda \in \mathcal{Q}$. Then the processes

$$[M + \lambda N, M + \lambda N] \quad \text{and} \quad [M, M] + 2\lambda[M, N] + \lambda^2[N, N]$$

are indistinguishable. Therefore, we can again apply theorem 2.20 according to which there exists an optional nonnegative process $Q(\lambda)$ such that

$$[M + \lambda N, M + \lambda N] = Q(\lambda) \cdot A$$

up to indistinguishability. Thus, the processes

$$\left(F + 2\lambda J + \lambda^2 G\right) \cdot A \quad \text{and} \quad Q(\lambda) \cdot A$$

are indistinguishable. Hence, P-a.s.,

$$Q(\lambda)_s = F_s + 2\lambda J_s + \lambda^2 G_s \qquad \text{for } dA_s\text{-almost all } s.$$

Consequently, with P-probability one,

$$F_s + 2\lambda J_s + \lambda^2 G_s \geqslant 0 \qquad \text{for } dA_s\text{-almost all } s.$$

holds simultaneously for all $\lambda \in \mathcal{Q}$ and, hence, simultaneously for all $\lambda \in \mathbb{R}$. Thus, P-a.s.

$$|J_s| \leqslant F_s^{1/2} G_s^{1/2} \qquad \text{for } dA_s\text{-almost all } t.$$

The claim follows from the Schwarz inequality:

$$\int_0^\infty |H_s K_s| \, d\operatorname{Var}[M, N]_s = \int_0^\infty |H_s K_s| |J_s| \, dA_s$$

$$\leqslant \int_0^\infty |H_s| F_s^{1/2} |K_s| G_s^{1/2} \, dA_s$$

$$\leqslant \left(\int_0^\infty H_s^2 F_s \, dA_s\right)^{1/2} \left(\int_0^\infty K_s^2 G_s \, dA_s\right)^{1/2}$$

$$= \left(\int_0^\infty H_s^2 \, d[M, M]_s\right)^{1/2} \left(\int_0^\infty K_s^2 \, d[N, N]_s\right)^{1/2}.$$

Inequality [2.53] is proved similarly. Inequalities [2.55] and [2.56] are obtained from [2.53] and [2.54], respectively, if we apply the Schwarz inequality.   □

We end this section with the remark that, using localization, the notions of (mutual) quadratic characteristic and quadratic (co)variation can be easily extended to *locally square integrable* martingales. The case of the quadratic variation will be considered in greater generality in the next section. Here we provide a necessary information concerning quadratic characteristics.

DEFINITION 2.23.– A local martingale M is called a *locally square-integrable martingale* if there is a localizing sequence $\{T_n\}$ of stopping times such that, for every n, we have $M^{T_n} \in \mathcal{M}^2$, i.e. the stopped process M^{T_n} is a square-integrable martingale. The class of all locally square-integrable martingales is denoted by $\mathcal{M}^2_{\mathrm{loc}}$. The *quadratic characteristic* of a locally square-integrable martingale $M \in \mathcal{M}^2_{\mathrm{loc}}$ is a predictable increasing process, denoted by $\langle M, M \rangle$ or $\langle M \rangle$, such that $M^2 - \langle M, M \rangle$ is a local martingale. The *mutual quadratic characteristic* of locally square-integrable martingales $M, N \in \mathcal{M}^2_{\mathrm{loc}}$ is a predictable process with finite variation, denoted by $\langle M, N \rangle$, such that $MN - \langle M, N \rangle$ is a local martingale.

The uniqueness (up to indistinguishability) of the quadratic characteristic and the mutual quadratic characteristic follow, as above, from corollary 2.14. Their existence is proved using the "gluing" procedure (proposition 2.9) similarly to the proof of the existence of the compensator in theorem 2.21.

For locally square-integrable martingales, basic properties of quadratic characteristics such as symmetry and bilinearity, lemma 2.15, theorem 2.26, the Kunita–Watanabe inequalities [2.53] and [2.55], still hold true.

EXERCISE 2.43.– Let $W = (W_t)_{t \in \mathbb{R}_+}$ be a Wiener process on a stochastic basis $(\Omega, \mathscr{F}, \mathbb{F}, \mathsf{P})$. Show that $W \in \mathcal{M}^2_{\mathrm{loc}}$, and find its quadratic characteristic $\langle W, W \rangle$.

2.8. Purely discontinuous local martingales

Unless otherwise stated, we will assume that a stochastic basis $\mathbb{B} = (\Omega, \mathscr{F}, \mathbb{F} = (\mathscr{F}_t)_{t \in \mathbb{R}_+}, \mathsf{P})$ satisfying the usual conditions is given.

It was shown in the previous section that square-integrable martingales admit a decomposition into the sum of continuous and purely discontinuous martingale components and the structure of purely discontinuous martingales was investigated. In this section, we fulfill a similar program for local martingales. Since stochastic processes are implicitly understood up to indistinguishability, it is natural to consider local martingales with a.s. continuous trajectories as continuous local martingales. With regard to purely discontinuous local martingales, our definition is motivated by the square-integrable case; see also lemma 2.21 below.

DEFINITION 2.24.– The space of *continuous local martingales* and the subspace of continuous local martingales starting from 0, are defined as

$$\mathcal{M}_{\text{loc}}^c := \{M \in \mathcal{M}_{\text{loc}} : M \text{ a.s. continuous}\}$$

$$\text{and } \mathcal{M}_{\text{loc},0}^c := \{N \in \mathcal{M}_{\text{loc}}^c : N_0 = 0 \text{ a.s.}\},$$

respectively. The space of *purely discontinuous local martingales* is defined as

$$\mathcal{M}_{\text{loc}}^d := \{M \in \mathcal{M}_{\text{loc}} : M_0 = 0 \text{ a.s. and } MN \in \mathcal{M}_{\text{loc}} \text{ for every } N \in \mathcal{M}_{\text{loc},0}^c\}.$$

Obviously, $\mathcal{M}_{\text{loc}}^c$, $\mathcal{M}_{\text{loc},0}^c$ and $\mathcal{M}_{\text{loc}}^d$ are linear spaces.

LEMMA 2.21.– Let $M \in \mathcal{M}_{\text{loc}}^d$. Then $MN \in \mathcal{M}_{\text{loc}}$ for every $N \in \mathcal{M}_{\text{loc}}^c$.

PROOF.– It is enough to check that, if $M \in \mathcal{M}_{\text{loc}}$, $M_0 = 0$ a.s., and ξ is an $\mathcal{F}_0$-measurable random variable, then $M\xi \in \mathcal{M}_{\text{loc}}$. Let $\{T_n\}$ be a localizing sequence for M, $S_n := 0_{\{|\xi| > n\}}$ is the restriction of the stopping time taking value zero identically on the set $\{|\xi| > n\}$. It is clear that $\{S_n\}$ and $\{S_n \wedge T_n\}$ are localizing sequences, $(M\xi)^{S_n \wedge T_n} = \xi M^{T_n} \mathbb{1}_{\{|\xi| \leqslant n\}}$, and it is easy to see that $\xi \mathbb{1}_{\{|\xi| \leqslant n\}} M^{T_n} \in \mathcal{M}$ for every n. □

REMARK 2.12.– If $M \in \mathcal{M}_{\text{loc},0}^c$ and $T_n := \inf\{t : |M_t| > n\}$, then M^{T_n} is a.s. bounded. In particular, $M^{T_n} \in \mathcal{M}^2$ and $M \in \mathcal{M}_{\text{loc}}^2$.

LEMMA 2.22.– The classes $\mathcal{M}_{\text{loc}}^c$ and $\mathcal{M}_{\text{loc}}^d$ are stable under stopping. If $M \in \mathcal{M}_{\text{loc}}$ and $M^{T_n} \in \mathcal{M}_{\text{loc}}^c$ (respectively $M^{T_n} \in \mathcal{M}_{\text{loc}}^d$) for some localizing sequence (T_n) of stopping times, then $M \in \mathcal{M}_{\text{loc}}^c$ (respectively $M \in \mathcal{M}_{\text{loc}}^d$).

PROOF.– The assertions concerning $\mathcal{M}_{\text{loc}}^c$ are obvious. If $M \in \mathcal{M}_{\text{loc}}$ and $M^{T_n} \in \mathcal{M}_{\text{loc}}^d$ for some localizing sequence (T_n) of stopping times, then, for every $N \in \mathcal{M}_{\text{loc},0}^c$, we have $M^{T_n} N \in \mathcal{M}_{\text{loc}}$. Hence, $M^{T_n} N^{T_n} \in \mathcal{M}_{\text{loc}}$ and $MN \in \mathcal{M}_{\text{loc}}$ by theorem 2.7.

Let $M \in \mathcal{M}_{\text{loc}}^d$, T be a stopping time, $N \in \mathcal{M}_{\text{loc},0}^c$. First let us prove that then $M^T N \in \mathcal{M}_{\text{loc}}$ under the following additional assumptions: $M^T \in \mathcal{H}^1$, N a.s. bounded. Since $M^T N = M^T N^T + M^T (N - N^T)$ and $M^T N^T = (MN)^T \in \mathcal{M}_{\text{loc}}$, it is enough to prove that the process $L := M^T (N - N^T)$ satisfies the assumptions of lemma 2.10, from which only the last one needs to be checked. Let S be an arbitrary stopping time. Then, due to the additional assumptions,

$$\mathsf{E}L_S = \mathsf{E}(M_{S \wedge T} N_S - M_{S \wedge T} N_{S \wedge T})$$

$$= \mathsf{E}(M_{S \wedge T} \mathsf{E}(N_S | \mathcal{F}_{S \wedge T}) - M_{S \wedge T} N_{S \wedge T}) = 0,$$

because $\mathsf{E}(N_S | \mathcal{F}_{S \wedge T}) = N_{S \wedge T}$ by theorem 2.4.

Now consider the general case. By theorem 2.7 and remark 2.12, there is a localizing sequence (T_n) of stopping times such that $M^{T_n} \in \mathscr{H}^1$ and N^{T_n} is a.s. bounded for every n. According to the previous case, $(M^T N)^{T_n} = M^{T \wedge T_n} N^{T_n} \in \mathscr{M}_{\text{loc}}$, hence $M^T N \in \mathscr{M}_{\text{loc}}$. Therefore, $M^T \in \mathscr{M}_{\text{loc}}^d$. $\square$

LEMMA 2.23.– $\mathscr{M}_{\text{loc}}^c \cap \mathscr{M}^2 = \mathscr{M}^{2,c}$, $\mathscr{M}_{\text{loc}}^d \cap \mathscr{M}^2 = \mathscr{M}^{2,d}$.

PROOF.– The first assertion is obvious. If $M \in \mathscr{M}_{\text{loc}}^d \cap \mathscr{M}^2$, then, for every $N \in \mathscr{M}^{2,c} \subseteq \mathscr{M}_{\text{loc}}^c$, we have $MN \in \mathscr{M}_{\text{loc}}$. But $\mathrm{E} M_\infty^* N_\infty^* < \infty$ due to [2.40], hence, $MN \in \mathscr{M}$ and $\mathrm{E} M_\infty N_\infty = 0$, i.e. $M \in \mathscr{M}^{2,d}$.

Conversely, let $M \in \mathscr{M}^{2,d}$ and $N \in \mathscr{M}_{\text{loc},0}^c$. By remark 2.12, $MN^{T_n} \in \mathscr{M}$, where $T_n := \inf\{t: |N_t| > n\}$. Therefore, $M^{T_n} N^{T_n} \in \mathscr{M}$ for every n, hence $MN \in \mathscr{M}_{\text{loc}}$ and $M \in \mathscr{M}_{\text{loc}}^d$. $\square$

LEMMA 2.24.– $\mathscr{M}_{\text{loc}} \cap \mathscr{V} \subseteq \mathscr{M}_{\text{loc}}^d$.

PROOF.– Let $M \in \mathscr{M}_{\text{loc}} \cap \mathscr{V} = \mathscr{M}_{\text{loc}} \cap \mathscr{A}_{\text{loc}}$ and $N \in \mathscr{M}_{\text{loc},0}^c$. Take a localizing sequence (T_n) of stopping times such that $M^{T_n} \in \mathscr{M}_{\text{loc}} \cap \mathscr{A}$ and N^{T_n} is a.s. bounded for every n. Then $M^{T_n} N^{T_n} \in \mathscr{M}$ by lemma 2.11. Hence, $MN \in \mathscr{M}_{\text{loc}}$ and $M \in \mathscr{M}_{\text{loc}}^d$. $\square$

EXERCISE 2.44.– Construct an example of a purely discontinuous local martingale starting from 0 for all ω, which does not belong to $\mathscr{V}$.

HINT.– Modify example 2.7: replace jumps times $t_n = n$ by $t_n = n/(n+1)$.

THEOREM 2.31.– Every local martingale admits a decomposition:

$$M = M_0 + M^c + M^d \qquad [2.57]$$

into the sum of a continuous local martingale $M^c \in \mathscr{M}_{\text{loc},0}^c$ and a purely discontinuous local martingale $M^d \in \mathscr{M}_{\text{loc}}^d$. This decomposition is unique up to indistinguishability.

REMARK 2.13.– If $M \in \mathscr{M}^2$, then decomposition [2.57] coincides with the decomposition into the continuous and purely discontinuous components, introduces in example 2.5. This follows from the uniqueness of decomposition [2.57] and lemma 2.23.

PROOF.– In order to prove the uniqueness, we have to check that the process $N \in \mathscr{M}_{\text{loc},0}^c \cap \mathscr{M}_{\text{loc}}^d$ vanishes except an evanescent set. By the definition of $\mathscr{M}_{\text{loc}}^d$, we have $N^2 \in \mathscr{M}_{\text{loc}}$. Being a nonnegative process, N^2 is a supermartingale by theorem 2.9. Hence, $0 \leqslant \mathrm{E} N_t^2 \leqslant \mathrm{E} N_0^2 = 0$, i.e., $N_t = 0$ a.s. for every $t \in \mathbb{R}_+$.

We turn to the proof of the existence. In view of the Gundy decomposition (theorem 2.22) and lemma 2.24, it is enough to consider the case where the jumps of M are bounded and $M_0 = 0$. Put $T_n := \inf\{t\colon |M_t| > n\}$, then M^{T_n} is bounded, in particular, $M^{T_n} \in \mathcal{M}^2$. It follows from section 2.7 that, for every n, there exists a unique decomposition

$$M^{T_n} = M^{n,c} + M^{n,d}, \quad M^{n,c} \in \mathcal{M}^{2,c}, \quad M^{n,d} \in \mathcal{M}^{2,d}.$$

Its uniqueness and lemma 2.22 imply that:

$M^{n,c}$ and $(M^{n,c})^{T_n}$ are indistinguishable for every $n = 1, 2, \ldots,$ [2.58]

$M^{n-1,c}$ and $(M^{n,c})^{T_{n-1}}$ are indistinguishable for every $n = 2, 3, \ldots.$ [2.59]

Let M^c be obtained by the "gluing" procedure from $\{M^{n,c}\}$, $\{T_n\}$. By proposition 2.9, the process M^c is a.s. continuous. Moreover,

$M^{n,c}$ and $(M^c)^{T_n}$ are indistinguishable for every $n = 1, 2, \ldots$

in view of [2.58] and [2.59]. Hence, putting $M^d := M - M^c$, we have that

$M^{n,d}$ and $(M^d)^{T_n}$ are indistinguishable for every $n = 1, 2, \ldots.$

Therefore, $M^c \in \mathcal{M}^c_{\mathrm{loc}}$ and $M^d \in \mathcal{M}^d_{\mathrm{loc}}$ by lemma 2.22.   □

LEMMA 2.25.– For every local martingale M and for every $t \in \mathbb{R}_+$,

$$\sum_{0 < s \leqslant t} (\Delta M_s)^2 < +\infty \quad \text{a.s.} \qquad [2.60]$$

PROOF.– Due to the Gundy decomposition (theorem 2.22) and inequality $(a + b)^2 \leqslant 2(a^2 + b^2)$, $a, b \in \mathbb{R}$, it is enough to consider two cases: $M \in \mathcal{M}^2_{\mathrm{loc}}$ and $M \in \mathcal{M}_{\mathrm{loc}} \cap \mathcal{V}$. In the first case, we may use relation [2.49] in theorem 2.29, which allow us to conclude that, for some localizing sequence (T_n),

$$\sum_{0 < s \leqslant T_n} (\Delta M_s)^2 < +\infty \quad \text{a.s.}$$

for every n, which, obviously, implies [2.60]. In the second case [2.60] follows from the inequality

$$\sum_{0 < s \leqslant t} (\Delta M_s)^2 \leqslant \left(\sum_{0 < s \leqslant t} |\Delta M_s| \right)^2 \leqslant \left(\mathrm{Var}\,(M)_t \right)^2. \qquad □$$

This lemma permits us to give the following definition.

DEFINITION 2.25.– The *quadratic variation* of a local martingale $M \in \mathcal{M}_{\mathrm{loc}}$ is the increasing process

$$[M, M] := \langle M^c, M^c \rangle + S\big((\Delta M)^2\big).$$

The *quadratic covariation* of local martingales $M, N \in \mathcal{M}_{\mathrm{loc}}$ is the process with finite variation

$$[M, N] := \langle M^c, N^c \rangle + S(\Delta M \Delta N).$$

Recall that $M^c, N^c \in \mathcal{M}^c_{\mathrm{loc},0} \subseteq \mathcal{M}^2_{\mathrm{loc}}$, so the angle brackets $\langle M^c, M^c \rangle$ and $\langle M^c, N^c \rangle$ are defined; see definition 2.23. As in the square-integrable case, the process $\langle M^c, N^c \rangle$ is a.s. continuous. Hence, the processes $\Delta[M, N]$ and $\Delta M \Delta N$ are indistinguishable for every $M, N \in \mathcal{M}_{\mathrm{loc}}$. Note also that in the case $M, N \in \mathcal{M}^2$ definition 2.25 coincides with definition 2.22; see remark 2.13.

Similarly to the square-integrable case, all relations between quadratic covariation processes are understood up to indistinguishability, the form $[M, N]$ is symmetric and bilinear, $[M, N^T] = [M, N]^T$ for every stopping time T and every $M, N \in \mathcal{M}_{\mathrm{loc}}$. Also, the Kunita–Watanabe inequalities [2.54] and [2.56] are still valid. It is less trivial to generalize the first assertion of lemma 2.20.

LEMMA 2.26.– If $M, N \in \mathcal{M}_{\mathrm{loc}}$ and $M_0 = N_0 = 0$ a.s., $MN - [M, N]$ is a local martingale.

PROOF.– Due to polarization, it is enough to consider the case $M = N$. In its turn, using the Gundy decomposition (theorem 2.22), we should check that:

$$M^2 - [M, M] \in \mathcal{M}_{\mathrm{loc}}, \tag{2.61}$$

$$MN - [M, N] \in \mathcal{M}_{\mathrm{loc}}, \tag{2.62}$$

$$N^2 - [N, N] \in \mathcal{M}_{\mathrm{loc}}, \tag{2.63}$$

where a local martingale M starts from 0 and has bounded jumps, and $N \in \mathcal{M}_{\mathrm{loc}} \cap \mathcal{V}$. Since $M \in \mathcal{M}^2_{\mathrm{loc}}$, [2.61] follows from lemma 2.20 by localization. Using localization again, we can assume that M is bounded and $N \in \mathcal{M}_{\mathrm{loc}} \cap \mathcal{A}$ in [2.62]. Then, by lemma 2.11, $MN - S(\Delta M \Delta N) \in \mathcal{M}$, and it remains to note that $N^c = 0$ by lemma 2.24. Finally, by Fubini's theorem, for every ω

$$
N_t^2 = \int_{]0,t]} dN_u \int_{]0,t]} dN_v = \int_{]0,t] \times]0,t]} dN_u dN_v
$$

$$
= \int_{]0,t] \times]0,t]} \left(\mathbb{1}_{\{u<v\}} + \mathbb{1}_{\{u>v\}} \right) dN_u dN_v + \int_{]0,t] \times]0,t]} \mathbb{1}_{\{u=v\}} dN_u dN_v
$$

$$
= 2 \int_{]0,t]} \left(\int_{]0,t]} \mathbb{1}_{\{u<v\}} dN_u \right) dN_v + \int_{]0,t]} \left(\int_{]0,t]} \mathbb{1}_{\{u=v\}} dN_u \right) dN_v
$$

$$
= 2 \int_{]0,t]} N_{v-} dN_v + \int_{]0,t]} \Delta N_v \, dN_v,
$$

i.e. $N^2 = 2N_- \cdot N + S\big((\Delta N)^2\big)$. Now [2.63] follows from proposition 2.10 (4). $\quad\square$

The following lemma strengthens lemma 2.25.

LEMMA 2.27.– Let M be a local martingale. Then $[M,M]^{1/2} \in \mathscr{A}_{\text{loc}}^+$.

PROOF.– In view of the inequality $(a+b)^{1/2} \leqslant a^{1/2} + b^{1/2}$, $a, b \in \mathbb{R}_+$, it is enough to consider separately the cases where M is either a continuous or a purely discontinuous local martingale.

In the first case, $[M,M] = \langle M,M \rangle$ is a.s. continuous and, hence, $[M,M]^{1/2} \in \mathscr{V}^+$ is a.s. continuous. Putting $T_n := \inf \{t \colon [M,M]_t^{1/2} > n\}$, we get $\mathsf{E}[M,M]_{T_n}^{1/2} \leqslant n$, i.e. $\big([M,M]^{1/2}\big)^{T_n} \in \mathscr{A}^+$.

In the second case, we have $[M,M] = S\big((\Delta M)^2\big)$. Using localization, we can assume that $M \in \mathcal{M}$. Put

$$
T_n := \inf \{t \colon S\big((\Delta M)^2\big)_t \geqslant n^2\} \wedge \inf \{t \colon |M_t| > n\}.
$$

Again using the elementary inequality from the beginning of the proof, we get

$$
\left\{ S\big((\Delta M)^2\big)_{T_n} \right\}^{1/2} \leqslant \left\{ S\big((\Delta M)^2\big)_{T_n-} \right\}^{1/2} + |\Delta M_{T_n}| \mathbb{1}_{\{T_n < \infty\}}
$$

$$
\leqslant 2n + |M_{T_n}| \mathbb{1}_{\{T_n < \infty\}}.
$$

Since the random variable on the right is integrable, the claim follows. $\quad\square$

One of the fundamental results of the theory of martingales is the Burkholder–Davis–Gundy inequality, which we formulate without proof. The case $p = 1$ is called Davis' inequality. Note that the previous lemma is a result of Davis' inequality and theorem 2.7.

THEOREM 2.32 (Burkholder–Davis–Gundy inequality).– Let M be a local martingale, $M_0 = 0$, T a stopping time, $p \geqslant 1$. There exist universal positive constants c_p and C_p (independent of T, M and a stochastic basis) such that

$$c_p \mathsf{E}[M, M]_T^{p/2} \leqslant \mathsf{E}(M_T^*)^p \leqslant C_p \mathsf{E}[M, M]_T^{p/2}. \qquad [2.64]$$

COROLLARY 2.18.– Let M be a local martingale, $M_0 = 0$, $p \geqslant 1$. Then

$$M \in \mathscr{H}^p \Leftrightarrow [M, M]^{p/2} \in \mathscr{A}^+, \qquad [2.65]$$

$$M \in \mathscr{H}_{\mathrm{loc}}^p \Leftrightarrow [M, M]^{p/2} \in \mathscr{A}_{\mathrm{loc}}^+. \qquad [2.66]$$

Here, $\mathscr{H}_{\mathrm{loc}}^p$ is the space of local martingales M such that there exists a localizing sequence $\{T_n\}$ with $M^{T_n} \in \mathscr{H}^p$ for every n.

THEOREM 2.33.– Let X be an optional process. There exists a (necessarily unique) purely discontinuous local martingale M such that

$$\Delta M = X$$

(up to indistinguishability) if and only if

$$\mathsf{E}(|X_T| \mathbb{1}_{\{T<\infty\}} | \mathscr{F}_{T-}) < \infty \quad \text{and} \quad \mathsf{E}(X_T \mathbb{1}_{\{T<\infty\}} | \mathscr{F}_{T-}) = 0 \qquad [2.67]$$

for every predictable stopping time T, the process $S(X^2)$ is well defined (see definition 2.21) and

$$\{S(X^2)\}^{1/2} \in \mathscr{A}_{\mathrm{loc}}^+. \qquad [2.68]$$

PROOF.– The necessity of conditions [2.67] and [2.68] follows from theorem 2.10 and lemma 2.27, and the uniqueness of M is a consequence of the uniqueness in theorem 2.31.

By theorem 1.17, there is a sequence (T_n) of stopping times such that

$$\{X \neq 0\} = \{\Delta S(X^2) \neq 0\} \subseteq \bigcup_n \llbracket T_n \rrbracket, \qquad [2.69]$$

(where the equality is up to evanescent set), every T_n is strictly positive and either predictable or totally inaccessible, and

$$[\![T_n]\!] \cap [\![T_m]\!] = \varnothing, \quad m \neq n. \tag{2.70}$$

Put $A^n = X_{T_n} \mathbb{1}_{\{T_n < \infty\}} \mathbb{1}_{[\![T_n, \infty[\![}}$. In view of [2.68], we have $A^n \in \mathscr{A}_{\mathrm{loc}}$, hence, the compensator $\widetilde{A}^n$ of A^n is defined. Put $M^n := A^n - \widetilde{A}^n$. By lemmas 2.8 and 2.9 (note that the assumptions of lemma 2.9 are valid because of [2.67]), we can assume that the processes $\widetilde{A}^n$ are continuous. Hence,

$$\Delta M^n = X \mathbb{1}_{[\![T_n]\!]}. \tag{2.71}$$

By lemma 2.24, $M^n \in \mathscr{M}_{\mathrm{loc}}^d$, hence, $[M^n, M^m] = 0$ for $m \neq n$ due to [2.70] and [2.71]. Put $N^n := M^1 + \cdots + M^n$, then

$$[N^n, N^n] = \sum_{j=1}^{n} [M^j, M^j] = S(X^2 \mathbb{1}_{\cup_{j=1}^{n} [\![T_j]\!]})$$

and

$$[N^{n+p} - N^n, N^{n+p} - N^n] = \sum_{j=n+1}^{n+p} [M^j, M^j] = S(X^2 \mathbb{1}_{\cup_{j=n+1}^{n+p} [\![T_j]\!]})$$

for every natural number p.

Let (S_k) be a localizing sequence such that $\mathsf{E}\{S(X^2)\}_{S_k}^{1/2} < \infty$ for every k; without loss of generality, $S_k \to \infty$ for all ω. By Davis' inequality,

$$\mathsf{E}(N^n)_{S_k}^* \leqslant C\mathsf{E}[N^n, N^n]_{S_k}^{1/2} \leqslant C\mathsf{E}\{S(X^2)\}_{S_k}^{1/2}$$

(in particular, $(N^n)^{S_k} \in \mathscr{H}^1$) and

$$\mathsf{E}(N^{n+p} - N^n)_{S_k}^* \leqslant C\mathsf{E}[N^{n+p} - N^n, N^{n+p} - N^n]_{S_k}^{1/2}$$

$$\leqslant C\mathsf{E}\{S(X^2 \mathbb{1}_{\cup_{j=n+1}^{\infty} [\![T_j]\!]})\}_{S_k}^{1/2}.$$

The expression on the right tends to 0 as $n \to \infty$ by the theorem on dominated convergence. Therefore, for every k, $(N^n)^{S_k}$ is a Cauchy sequence in $\mathscr{H}^1$ and, hence, by theorem 2.6, converges in $\mathscr{H}^1$ to a limit, which we denote by L^k. It is easy to see that $L^k = (L^k)^{S_k}$ and $L^{k-1} = (L^k)^{S_{k-1}}$. So, using the "gluing" procedure, we can construct a process $L \in \mathscr{H}^1_{\mathrm{loc}} = \mathscr{M}_{\mathrm{loc}}$ such that $L^{S_k} = L^k$ for every k (all equalities are up to indistinguishability). Now note that, by the construction, for every k,

$$\mathsf{E} \sup_{s \leqslant S_k} |N^n_s - L_s| \to 0.$$

On the other hand, for every $t \in \mathbb{R}_+$,

$$\mathsf{P}\Big(\sup_{s \leqslant t} |N^n_s - L_s| > \varepsilon\Big) \leqslant \mathsf{P}\Big(\sup_{s \leqslant S_k} |N^n_s - L_s| > \varepsilon\Big) + \mathsf{P}(S_k < t).$$

Passing to the limit as $n \to \infty$ and then as $k \to \infty$, we get

$$\sup_{s \leqslant t} |N^n_s - L_s| \xrightarrow{P} 0.$$

Choosing a subsequence converging almost surely and using [2.69]–[2.71], we obtain that the processes ΔL and X are indistinguishable. It remains to set $M := L^d$. $\square$

EXERCISE 2.45.– Show that the process L in the proof of theorem 2.33 is itself a purely discontinuous martingale.

Stochastic Integrals

Unless otherwise stated, we will assume in this chapter that a stochastic basis $\mathbb{B} = (\Omega, \mathscr{F}, \mathbb{F} = (\mathscr{F}_t)_{t \in \mathbb{R}_+}, \mathsf{P})$ satisfying the usual conditions is given.

3.1. Stochastic integrals with respect to local martingales

The purpose of this section is to define a stochastic integral process $\int_{]0,t]} H_s \, dM_s$ for a local martingale $M = (M_t)$ for a wide class of possible predictable integrands $H = (H_t)$. Here we limit ourselves to those H, for which the integral process is a local martingale (we know from example 2.3 that this is not always the case). At least continuous martingale are not processes with finite variations (corollary 2.14), therefore, an integral with respect to such processes cannot be defined as a pathwise Lebesgue–Stieltjes integral. The construction introduced below is called the *stochastic integral*. The integral process, as in the case of the Lebesgue–Stieltjes integral, will be denoted by $H \cdot M = (H \cdot M_t)$. To distinguish from the stochastic integral, the pathwise Lebesgue–Stieltjes integral (which is defined in theorem 2.11) will be denoted by $H \overset{s}{\cdot} M$ (if $M \in \mathscr{V}$).

First, let us define the stochastic integral $H \cdot M$ for square-integrable martingales M and for predictable H, such that the integral process $H \cdot M$ is again a square-integrable martingale. For this purpose, for a given $M \in \mathscr{M}^2$, we introduce the space

$$
L^2(M) := \left\{ H \text{ predictable} : \mathsf{E} \int_0^\infty H_s^2 \, d\langle M, M \rangle_s < \infty \right\}
$$

$$
= \left\{ H \text{ predictable} : \int_{\Omega \times \mathbb{R}_+} H^2 \, d\mu_{\langle M, M \rangle} < \infty \right\},
$$

where $\mu_{\langle M,M \rangle}$ is the Doléans measure of the integrable increasing process $\langle M, M \rangle$. In view of the last equality, $L^2(M)$ is none other than the space L^2 of measurable square-integrable functions on the measurable space $(\Omega \times \mathbb{R}_+, \mathscr{P}, \mu_{\langle M,M \rangle}|_{\mathscr{P}})$.

Let us also introduce the class Λ of ("simple") predictable processes of the form:

$$H = \eta \mathbb{1}_{[0]} + \sum_{i=1}^{n} \xi_i \mathbb{1}_{]\!]t_{i-1},t_i]\!]},$$ [3.1]

where η is a bounded $\mathscr{F}_0$-measurable random variable, $n = 0, 1, \ldots, 0 = t_0 < t_1 < \cdots < t_n < +\infty$, ξ_i $(i = 1, \ldots, n)$ are bounded $\mathscr{F}_{t_{i-1}}$-measurable random variables. It is clear that $\Lambda \subseteq L^2(M)$ for every $M \in \mathscr{M}^2$. For $H \in \Lambda$ of the form [3.1] and for $M \in \mathscr{M}^2$, let us define the process $H \cdot M = (H \cdot M_t)$ by

$$H \cdot M_t := \sum_{i=1}^{n} \xi_i \big(M_{t_i \wedge t} - M_{t_{i-1} \wedge t} \big).$$

LEMMA 3.1.– Let $H \in \Lambda$ and $M \in \mathscr{M}^2$. Then, $H \cdot M \in \mathscr{M}^2$ and

$$\mathsf{E}(H \cdot M_\infty)^2 = \mathsf{E} \int_0^\infty H_s^2 \, d\langle M, M \rangle_s.$$

If $N \in \mathscr{M}^2$, then

$$\mathsf{E}(H \cdot M_\infty) N_\infty = \mathsf{E} \int_0^\infty H_s \, d\langle M, N \rangle_s.$$

PROOF.– Obviously, $H \cdot M$ is a martingale. Since $H \cdot M_t = H \cdot M_{t_n}$ for $t \geqslant t_n$, we have $H \cdot M \in \mathscr{M}$. Furthermore, let $1 \leqslant i < j \leqslant n$. Then $i \leqslant j - 1$ and

$$\mathsf{E}\xi_i\xi_j \big(M_{t_i \wedge t} - M_{t_{i-1} \wedge t} \big) \big(M_{t_j \wedge t} - M_{t_{j-1} \wedge t} \big)$$
$$= \mathsf{E}\mathsf{E}\Big[\xi_i\xi_j \big(M_{t_i \wedge t} - M_{t_{i-1} \wedge t} \big) \big(M_{t_j \wedge t} - M_{t_{j-1} \wedge t} \big) \Big| \mathscr{F}_{t_{j-1}} \Big]$$
$$= \mathsf{E}\xi_i\xi_j \big(M_{t_i \wedge t} - M_{t_{i-1} \wedge t} \big) \mathsf{E}\big(M_{t_j \wedge t} - M_{t_{j-1} \wedge t} \big| \mathscr{F}_{t_{j-1}} \big) = 0.$$

Hence,

$$E(H \cdot M_\infty)^2 = E \sum_{i=1}^{n} \xi_i^2 (M_{t_i} - M_{t_{i-1}})^2 = E \sum_{i=1}^{n} \xi_i^2 (M_{t_i}^2 - M_{t_{i-1}}^2)$$

$$= E \sum_{i=1}^{n} \xi_i^2 (\langle M \rangle_{t_i} - \langle M \rangle_{t_{i-1}}) = E \int_0^\infty H_s^2 \, d\langle M, M \rangle_s$$

(to prove the second equality, we have used exercise 2.41). If $N \in \mathcal{M}^2$, we have

$$E\xi_i (M_{t_i} - M_{t_{i-1}}) N_\infty = E\xi_i (M_{t_i} - M_{t_{i-1}}) N_{t_i}$$

and

$$E\xi_i (M_{t_i} - M_{t_{i-1}}) N_{t_{i-1}} = 0.$$

Similarly, we get

$$E(H \cdot M_\infty) N_\infty = E \sum_{i=1}^{n} \xi_i (M_{t_i} - M_{t_{i-1}}) N_\infty$$

$$= E \sum_{i=1}^{n} \xi_i (M_{t_i} - M_{t_{i-1}})(N_{t_i} - N_{t_{i-1}})$$

$$= E \sum_{i=1}^{n} \xi_i (M_{t_i} N_{t_i} - M_{t_{i-1}} N_{t_{i-1}})$$

$$= E \sum_{i=1}^{n} \xi_i (\langle M, N \rangle_{t_i} - \langle M, N \rangle_{t_{i-1}}) = E \int_0^\infty H_s \, d\langle M, N \rangle_s. \qquad \square$$

THEOREM 3.1.– Let $M \in \mathcal{M}^2$. The mapping $H \rightsquigarrow H \cdot M$ from Λ to $\mathcal{M}^2$ extends uniquely to a linear isometry of $L^2(M)$ into $\mathcal{M}^2$. This extension is also denoted by $H \rightsquigarrow H \cdot M$. Moreover, $H \cdot M_0 = 0$ and $\Delta(H \cdot M) = H \Delta M$ (up to an evanescent set) for every $H \in L^2(M)$.

PROOF.– Obviously, the mapping $H \rightsquigarrow H \cdot M$ from Λ to $\mathcal{M}^2$ is linear. It is also an isometry by lemma 3.1. Hence, to prove the first assertion, it is enough to show that Λ is dense in $L^2(M)$. The last statement follows from the facts that in any L^2-space (with respect to a finite measure), an arbitrary function can be approximated by bounded ones, the bounded function can be approximated by a finite-valued function, and an indicator of a set can be approximated by an indicator of a set from a given algebra that generates the σ-algebra, and from theorem 1.10.

Since $H \cdot M_0 = 0$ is true for $H \in \Lambda$, it holds for every $H \in L^2(M)$.

It is clear that $\Delta(H \cdot M) = H\Delta M$ for all $H \in \Lambda$. Let $\mathscr{H}$ be the class of those bounded predictable processes H for which this equality holds (up to an evanescent set). Since the mapping $H \rightsquigarrow H \cdot M$ is linear, it is obvious that $\mathscr{H}$ is a linear space containing constants. Let a uniformly bounded sequence $\{H^n\}$ in $\mathscr{H}$ converges to H for all (ω, t). Then, on the one hand,

$$\Delta(H^n \cdot M) = H^n \Delta M \to H\Delta M.$$

On the other hand, H^n converges to H in $L^2(M)$. Therefore, $H^n \cdot M$ converges to $H \cdot M$ in $\mathscr{M}^2$ due to the isometry property. By lemma 2.14, there is a subsequence n_k such that, for a.a. ω,

$$\Delta(H^{n_k} \cdot M) \to \Delta(H \cdot M).$$

Therefore, $H\Delta M$ and $\Delta(H \cdot M)$ are indistinguishable. A monotone class argument (theorem A.3) and theorem 1.10 imply that $\mathscr{H}$ contains all bounded predictable processes.

If H is an arbitrary predictable process in $L^2(M)$, then the equality $H\Delta M = \Delta(H \cdot M)$ is proved quite similar to the above by passing to the limit from $H^n = H\mathbb{1}_{\{|H| \leqslant n\}}$.   □

REMARK 3.1.– It is essential for the above argument that H^n converges to H pointwise. If H^n converges to H only in $L^2(M)$, then we can assert, passing to a subsequence, that $H_t^{n_k}(\omega) \to H_t(\omega)$ for $\mu_{\langle M,M \rangle}$-a.a. (ω, t). It does not follow directly (but can be proved) from the last relation that for almost all ω $H_t^{n_k}(\omega)\Delta M_t(\omega) \to H_t(\omega)\Delta M_t(\omega)$ for all t.

DEFINITION 3.1.– Let $M \in \mathscr{M}^2$ and $H \in L^2(M)$. The process $H \cdot M$ determined in theorem 3.1 is called the *stochastic integral* of H with respect to M.

The stochastic integral process is determined up to an evanescent set and does not depend on the choice of versions of H and M. The same refers to other constructions of stochastic integrals presented below in this chapter. Thus, all pointwise relations containing stochastic integral processes are understood only up to an evanescent set.

EXERCISE 3.1.– Let $M \in \mathscr{M}^2$ and T be a stopping time.

1) Show that

$$\mathbb{1}_{]\!]0,T]\!]} \cdot M = \mathbb{1}_{[\![0,T]\!]} \cdot M = M^T - M_0.$$

2) Let T be predictable. Find $\mathbb{1}_{]0,T[} \cdot M$ and $\mathbb{1}_{[0,T[} \cdot M$.

The next theorem provides a useful characterization of the stochastic integral.

THEOREM 3.2.– Let $M \in \mathcal{M}^2$ and $H \in L^2(M)$. Then, for every $N \in \mathcal{M}^2$,

$$\mathsf{E} \int_0^\infty |H_t| \, d\operatorname{Var}\left(\langle M, N \rangle\right)_t < \infty, \qquad [3.2]$$

$$\mathsf{E}(H \cdot M_\infty) N_\infty = \mathsf{E} H \overset{s}{\cdot} \langle M, N \rangle_\infty, \qquad [3.3]$$

$$\langle H \cdot M, N \rangle = H \overset{s}{\cdot} \langle M, N \rangle. \qquad [3.4]$$

If $L \in \mathcal{M}^2$ and $\mathsf{E} L_\infty N_\infty$ coincides with the expression on the right in [3.3] for every $N \in \mathcal{M}^2$, then $L = H \cdot M$. If $L \in \mathcal{M}^2$, $L_0 = 0$ and $\langle L, N \rangle = H \overset{s}{\cdot} \langle M, N \rangle$ for every $N \in \mathcal{M}^2$, then $L = H \cdot M$.

PROOF.– [3.2] follows directly from the Kunita–Watanabe inequality [2.55]. For fixed $M, N \in \mathcal{M}^2$, consider the functional

$$H \rightsquigarrow \mathsf{E}\left[(H \cdot M_\infty) N_\infty - H \overset{s}{\cdot} \langle M, N \rangle_\infty\right]$$

on $L^2(M)$. It is linear and continuous (take [2.55] into account). By lemma 3.1, it vanishes on Λ, and it was shown in the proof of theorem 3.1 that Λ is dense in $L^2(M)$. It follows that the functional vanishes identically, which proves [3.3].

The process $H \overset{s}{\cdot} \langle M, N \rangle$ is predictable and belongs to $\mathscr{A}$ due to [3.2]. Hence, to prove [3.4], it is enough to check that $(H \cdot M)N - H \overset{s}{\cdot} \langle M, N \rangle \in \mathcal{M}$. Let T be a stopping time. Then

$$\mathsf{E}(H \overset{s}{\cdot} \langle M, N \rangle_T) = \mathsf{E}(H \overset{s}{\cdot} \langle M, N^T \rangle)_\infty = \mathsf{E}(H \cdot M_\infty) N_T = \mathsf{E}(H \cdot M_T) N_T,$$

where we make use of [3.3] in the second equality. Lemma 2.10 yields the claim.

Last two assertions of the theorem are obvious.   □

COROLLARY 3.1.– Let $M \in \mathcal{M}^2$, $H \in L^2(M)$, and T be a stopping time. Then

$$(H \cdot M)^T = H \cdot M^T = (H\mathbb{1}_{]0,T]}) \cdot M.$$

PROOF.– The assertion follows directly from theorem 3.2.   □

COROLLARY 3.2.– Let $M, N \in \mathcal{M}^2$, $H \in L^2(M) \cap L^2(N)$, $\alpha, \beta \in \mathbb{R}$. Then, $H \in L^2(\alpha M + \beta N)$ and $H \cdot (\alpha M + \beta N) = \alpha(H \cdot M) + \beta(H \cdot N)$.

PROOF.– It follows from the Kunita–Watanabe inequality [2.55] with $K = H$ that $H \in L^2(\alpha M + \beta N)$. Now the equality $H \cdot (\alpha M + \beta N) = \alpha(H \cdot M) + \beta(H \cdot N)$ follows from theorem 3.2. $\qquad\square$

COROLLARY 3.3.– Let $\mathcal{H}$ be a stable subspace of $\mathcal{M}^2$, $M \in \mathcal{H}$, $H \in L^2(M)$. Then, $H \cdot M \in \mathcal{H}$.

PROOF.– Let $N \in \mathcal{H}^\perp$. By theorem 2.27, M and N are strongly orthogonal. Hence, $\langle M, N \rangle = 0$ by lemma 2.16. Theorem 3.2 implies $\langle H \cdot M, N \rangle = 0$, hence, by lemma 2.16, $H \cdot M$ and N are strongly orthogonal and $H \cdot M \in \mathcal{H}$. $\qquad\square$

COROLLARY 3.4.– Let $M \in \mathcal{M}^2$ and $H \in L^2(M)$. Then, $H \in L^2(M^c) \cap L^2(M^d)$, $(H \cdot M)^c = H \cdot M^c$, $(H \cdot M)^d = H \cdot M^d$.

PROOF.– The first assertion follows from the equality $\langle M, M \rangle = \langle M^c, M^c \rangle + \langle M^d, M^d \rangle$. By corollary 3.2, $H \cdot M = H \cdot M^c + H \cdot M^d$, and $H \cdot M^c \in \mathcal{M}^{2,c}$ and $H \cdot M^d \in \mathcal{M}^{2,d}$ by corollary 3.3. $\qquad\square$

COROLLARY 3.5.– Let $M \in \mathcal{M}^2$, $M_0 = 0$. Then the smallest stable subspace containing M is $\mathcal{H} := \{H \cdot M : H \in L^2(M)\}$. Let $N \in \mathcal{M}^2$. There is a process K in $L^2(M)$ such that $\langle M, N \rangle = K \overset{s}{\cdot} \langle M, M \rangle$. For any K with these properties, the projection N onto $\mathcal{H}$ is $K \cdot M$.

PROOF.– Since $M_0 = 0$, we have $M = 1 \cdot M \in \mathcal{H}$. By corollary 3.3, any stable subspace containing M, contains $\mathcal{H}$ as well. Therefore, to prove the first assertion, it is enough to show that $\mathcal{H}$ is a stable subspace. Since the mapping $H \leadsto H \cdot M$ from $L^2(M)$ into $\mathcal{M}^2$ is a linear isometry, $\mathcal{H}$ is a norm-closed linear subspace of $\mathcal{M}^2$. Next, it follows from corollary 3.1 that $\mathcal{H}$ is stable under stopping. The last property from the definition of a stable subspace is trivially true. Hence, $\mathcal{H}$ is a stable subspace.

Let $N \in \mathcal{M}^2$. As we have just proved, the projection of N onto $\mathcal{H}$ has the form $K \cdot M$ for some $K \in L^2(M)$. Since $N - K \cdot M \in \mathcal{H}^\perp$, we have $\langle N - K \cdot M, M \rangle = 0$, hence $\langle M, N \rangle = K \overset{s}{\cdot} \langle M, M \rangle$ in view of [3.4]. Conversely, if $K \in L^2(M)$ and $\langle M, N \rangle = K \overset{s}{\cdot} \langle M, M \rangle$, then $\langle N - K \cdot M, M \rangle = 0$ and $\langle N - K \cdot M, H \cdot M \rangle = H \overset{s}{\cdot} \langle N - K \cdot M, M \rangle = 0$ due to [3.4]. Thus, $N - K \cdot M \in \mathcal{H}^\perp$. $\qquad\square$

COROLLARY 3.6.– Let $M \in \mathcal{M}^2$, $H \in L^2(M)$. Then, for every $N \in \mathcal{M}^2$,

$$\mathbb{E} \int_0^\infty |H_t| \, d\operatorname{Var}([M, N])_t < \infty \quad \text{and} \quad [H \cdot M, N] = H \overset{s}{\cdot} [M, N].$$

If $L \in \mathcal{M}^2$, $L_0 = 0$ and $[L, N] = H \overset{s}{\cdot} [M, N]$ for every $N \in \mathcal{M}^2$, then $L = H \cdot M$.

PROOF.– The first assertion follows from the Kunita–Watanabe inequality [2.56], the definition of quadratic covariation, corollary 3.4, [3.4], and theorem 3.1. The second assertion reduces to the final statement in theorem 3.2 with the use of lemma 2.20 and theorem 2.21 (2). □

COROLLARY 3.7.– Let $M \in \mathcal{M}^2$, $H \in L^2(M)$, and K be a predictable process. Then $K \in L^2(H \cdot M)$ if and only if $KH \in L^2(M)$. In this case, $K \cdot (H \cdot M) = (KH) \cdot M$.

PROOF.– The assertion follows from theorem 3.2 and a similar property of the Lebesgue–Stieltjes integral, see theorem 2.12 (4). □

THEOREM 3.3.– Let $M \in \mathcal{M}^2 \cap \mathcal{V}$, $H \in L^2(M) \cap L_{var}(M)$ and $H \overset{s}{\cdot} M \in \mathcal{A}_{loc}$. Then $H \cdot M = H \overset{s}{\cdot} M$.

PROOF.– By lemmas 2.24 and 2.23, $M \in \mathcal{M}^{2,d}$. Hence, by corollary 3.3, $H \cdot M \in \mathcal{M}^{2,d} \subseteq \mathcal{M}^d_{loc}$. On the other hand, $H \overset{s}{\cdot} M \in \mathcal{M}_{loc} \cap \mathcal{A}_{loc} \subseteq \mathcal{M}^d_{loc}$ by proposition 2.10 (4) and lemma 2.24. It remains to note that, by theorem 3.1, $\Delta(H \cdot M) = H\Delta M = \Delta(H \overset{s}{\cdot} M)$. □

Taking corollary 3.5 into account, we may conjecture that the smallest stable subspace containing a finite number of square-integrable martingales $M^1, \dots, M^n$, where $M^1_0 = \cdots = M^n_0 = 0$, consists of sums $\{H^1 \cdot M^1 + \cdots + H^n \cdot M^n : H^1 \in L^2(M^1), \dots, H^n \in L^2(M^n)\}$. The following example shows that it is not the case, in general. Namely, the subspace of sums is not necessarily norm-closed. In fact, the smallest stable subspace containing $M^1, \dots, M^n$, is the closure of the linear subspace $\{H^1 \cdot M^1 + \cdots + H^n \cdot M^n : H^1 \in L^2(M^1), \dots, H^n \in L^2(M^n)\}$ in $\mathcal{M}^2$. It can be also described with the use of so-called *vector* stochastic integral, see [SHI 02].

EXAMPLE 3.1.– Put $A_t := t \wedge 1$, $t \in \mathbb{R}_+$. Let M and $\widetilde{M}$ be martingales from $\mathcal{M}^2$ such that $M_0 = \widetilde{M}_0 = 0$, $\langle M \rangle = \langle \widetilde{M} \rangle = A$ and $\langle M, \widetilde{M} \rangle = 0$, see exercise 3.2 concerning the existence of such processes. Let us also take a measurable function K_s with values in $(0, 1)$. We can consider $K = (K_s)$ as a (deterministic) bounded predictable stochastic process. Thus, we can define:

$$N := K \cdot M + (1 - K) \cdot \widetilde{M} \in \mathcal{M}^2.$$

We will show that $\widetilde{M}$ always belongs to the smallest stable subspace $\mathcal{H}$ containing M and N, while a representation $\widetilde{M} = H \cdot M + G \cdot N$ with $H \in L^2(M)$, $G \in L^2(N)$, exists only under an additional assumption on K.

Put

$$H^n := -K \mathbb{1}_{\{K \leqslant 1 - \frac{1}{n}\}}/(1-K), \quad G^n := \mathbb{1}_{\{K \leqslant 1 - \frac{1}{n}\}}/(1-K).$$

Then, H^n and G^n are bounded and predictable (deterministic, in fact) stochastic processes, and we can define

$$\widetilde{M}^n := H^n \cdot M + G^n \cdot N,$$

moreover, $\widetilde{M}^n \in \mathscr{H}$ by corollary 3.3. Due to corollary 3.7, we get

$$\widetilde{M}^n = H^n \cdot M + (G^n K) \cdot M + (G^n(1-K)) \cdot \widetilde{M} = \mathbb{1}_{\{K \leqslant 1 - \frac{1}{n}\}} \cdot \widetilde{M}.$$

Owing to [3.4],

$$\langle \widetilde{M} - \widetilde{M}^n, \widetilde{M} - \widetilde{M}^n \rangle = \langle \mathbb{1}_{\{K > 1 - \frac{1}{n}\}} \cdot \widetilde{M}, \mathbb{1}_{\{K > 1 - \frac{1}{n}\}} \cdot \widetilde{M} \rangle = \mathbb{1}_{\{K > 1 - \frac{1}{n}\}} \stackrel{s}{\cdot} \langle \widetilde{M}, \widetilde{M} \rangle$$

and

$$\mathsf{E}(\widetilde{M}_\infty - \widetilde{M}^n_\infty)^2 = \mathsf{E}\langle \widetilde{M} - \widetilde{M}^n, \widetilde{M} - \widetilde{M}^n \rangle_\infty = \mathsf{E}\int_0^1 \mathbb{1}_{\{K_t > 1 - \frac{1}{n}\}}\, dt \to 0$$

as $n \to \infty$. Thus, $\widetilde{M} \in \mathscr{H}$.

Assume now that $\widetilde{M} = H \cdot M + G \cdot N$, where $H \in L^2(M)$ and $G \in L^2(N)$. By corollary 3.7,

$$\widetilde{M} = (H + GK) \cdot M + (G(1-K)) \cdot \widetilde{M},$$

hence

$$(H + GK) \cdot M + (G(1-K) - 1) \cdot \widetilde{M} = 0.$$

Taking into account [3.4], we have

$$
\begin{aligned}
0 &= \langle (H + GK) \cdot M + (G(1-K) - 1) \cdot \widetilde{M} \rangle \\
&= (H + GK)^2 \stackrel{s}{\cdot} \langle M, M \rangle + [2(H + GK)(G(1-K) - 1)] \stackrel{s}{\cdot} \langle M, \widetilde{M} \rangle \\
&\quad + (G(1-K) - 1)^2 \stackrel{s}{\cdot} \langle \widetilde{M}, \widetilde{M} \rangle \\
&= [(H + GK)^2 + (G(1-K) - 1)^2] \stackrel{s}{\cdot} A.
\end{aligned}
$$

In particular,

$$G_t = \frac{1}{1 - K_t} \quad \text{i} \quad H_t = -\frac{K_t}{1 - K_t} \quad dP\, dA_t\text{-a.e.}$$

The condition $H \in L^2(M)$ implies:

$$\int_0^1 \left(\frac{K_t}{1 - K_t} \right)^2 dt < \infty.$$

This is not always true. For example, take $K_s = 1 - s$ for $s \in (0, 1/2)$ and $K_s = 1/2$ for other s.

EXERCISE 3.2.– Construct martingales M and $\widetilde{M}$, satisfying the assumptions in example 3.1.

HINT.– Take two independent Wiener processes W and $\widetilde{W}$ and stop them at time 1.

The next step in defining a stochastic integral is to localize definition 3.1. To this aim, given $M \in \mathcal{M}_{\mathrm{loc}}^2$, we introduce a space

$$L_{\mathrm{loc}}^2(M) := \left\{ H \text{ predictable} : H^2 \in L_{\mathrm{var}}(\langle M, M \rangle) \right\}$$

$$= \left\{ H \text{ predictable} : \int_0^t H_s^2\, d\langle M, M \rangle_s < \infty \text{ P-a.s. for every } t \in \mathbb{R}_+ \right\}.$$

DEFINITION 3.2.– Let $M \in \mathcal{M}_{\mathrm{loc}}^2$ and $H \in L_{\mathrm{loc}}^2(M)$. *The stochastic integral* process of H with respect to M is a locally square-integrable martingale N such that the following is true: if T is a stopping time such that $M^T \in \mathcal{M}^2$ and $H \in L^2(M^T)$, then

$$N^T = H \cdot M^T.$$

The stochastic integral process will be denoted by $H \cdot M$.

PROPOSITION 3.1.– The process N with the properties as in definition 3.2 exists and is unique (up to an evanescent set).

PROOF.– Let $M \in \mathcal{M}_{\mathrm{loc}}^2$ and $H \in L_{\mathrm{loc}}^2(M)$. Since $H^2 \overset{s}{\cdot} \langle M, M \rangle$ is a predictable process, it belongs to $\mathcal{A}_{\mathrm{loc}}^+$ by lemma 2.5. Hence, there is a localizing sequence $\{T_n\}$ such that

$$M^{T_n} \in \mathcal{M}^2 \quad \text{and} \quad H^2 \overset{s}{\cdot} \langle M, M \rangle^{T_n} \in \mathcal{A}^+ \quad \text{for every } n.$$

In particular,

$$H \in L^2(M^{T_n}) \quad \text{for every } n.$$

The uniqueness follows. Next, by corollary 3.1,

$$H \cdot M^{T_n} \text{ and } \left(H \cdot M^{T_n}\right)^{T_n} \text{ are indistinguishable for every } n = 1, 2, \ldots,$$

and

$$H \cdot M^{T_{n-1}} \text{ and } \left(H \cdot M^{T_n}\right)^{T_{n-1}} \text{ are indistinguishable for every } n = 2, 3, \ldots.$$

The "gluing" procedure yields an adapted càdlàg process N such that:

$$H \cdot M^{T_n} \text{ and } N^{T_n} \text{ are indistinguishable for every } n = 1, 2, \ldots.$$

Since $H \cdot M^{T_n} \in \mathcal{M}^2$, we have $N \in \mathcal{M}_{\text{loc}}^2$.

Let now T be a stopping time, $M^T \in \mathcal{M}^2$ and $H \in L^2(M^T)$. Then $M^{T \wedge T_n} \in \mathcal{M}^2$ and $H \in L^2(M^{T \wedge T_n})$ for every n. By construction and using corollary 3.1, we get:

$$N^{T \wedge T_n} = \left(H \cdot M^{T_n}\right)^T = H \cdot M^{T \wedge T_n} = \left(H \cdot M^T\right)^{T_n}.$$

Hence, $N^T = H \cdot M^T$. Thus, the existence of N from definition 3.2 is also proved.

$\square$

Note that the process $H \cdot M$ from definition 3.2 coincides with the previously introduced in theorem 3.1 process $H \cdot M$ in the case $M \in \mathcal{M}^2$, $H \in L^2(M)$. So our notation is not ambiguous.

It is easy to verify that the stochastic integral $H \cdot M$ introduced in definition 3.2 is linear with respect to each of two arguments, and we have, for every $M \in \mathcal{M}_{\text{loc}}^2$ and $H \in L_{\text{loc}}^2(M)$,

– $H \cdot M_0 = 0$ and $\Delta(H \cdot M) = H \Delta M$ (up to an evanescent set);

– for every $N \in \mathcal{M}_{\text{loc}}^2$, the processes $H \overset{s}{\cdot} \langle M, N \rangle$, $H \overset{s}{\cdot} [M, N]$ are well defined and $\langle H \cdot M, N \rangle = H \overset{s}{\cdot} \langle M, N \rangle$, $[H \cdot M, N] = H \overset{s}{\cdot} [M, N]$;

– $(H \cdot M)^T = H \cdot M^T = (H \mathbb{1}_{]0,T]}) \cdot M$ for every stopping time T;

– $H \in L_{\text{loc}}^2(M^c) \cap L_{\text{loc}}^2(M^d)$, $(H \cdot M)^c = H \cdot M^c$, $(H \cdot M)^d = H \cdot M^d$;

– if K is a predictable process, then $K \in L^2_{\text{loc}}(H \cdot M)$ if and only if $KH \in L^2_{\text{loc}}(M)$. In this case, $K \cdot (H \cdot M) = (KH) \cdot M$;

– if, additionally, $M \in \mathscr{V}$, $H \in L_{\text{var}}(M)$, and $H \overset{s}{\cdot} M \in \mathscr{A}_{\text{loc}}$, then $H \cdot M = H \overset{s}{\cdot} M$.

EXERCISE 3.3.– Prove the above assertions.

It is also useful to note that, for $M \in \mathscr{M}^2_{\text{loc}}$, the angle bracket $\langle M, M \rangle$ is the compensator of the square bracket $[M, M]$, see lemma 2.26. Therefore, by theorem 2.21, we can replace $\langle M, M \rangle$ by $[M, M]$ in the definition of $L^2(M)$ and $L^2_{\text{loc}}(M)$.

We now turn to the definition of the stochastic integral with respect to an arbitrary local martingale M. One possibility is to use the Gundy decomposition (theorem 2.22) $M = M_0 + M^1 + M^2$ into the sum of a local martingale M^1 with bounded jumps (and, hence, locally square-integrable) and a local martingale with bounded variation M^2, and to define $H \cdot M$ as $H \cdot M^1 + H \overset{s}{\cdot} M^2$. Though this approach is natural, we use the other approach based on the decomposition [2.57] of a local martingale into the sum of the continuous and purely discontinuous martingale components and on theorem 2.33, characterizing the jumps of purely discontinuous martingales.

Let $M \in \mathscr{M}_{\text{loc}}$ and $p \geqslant 1$. Define the classes of integrands $L^p(M)$ and $L^p_{\text{loc}}(M)$ by

$$L^p(M) := \left\{ H \text{ predictable} \colon H^2 \in L_{\text{var}}([M, M]) \text{ and } \left(H^2 \overset{s}{\cdot} [M, M] \right)^{p/2} \in \mathscr{A}^+ \right\},$$

$$L^p_{\text{loc}}(M) := \left\{ H \text{ predictable} \colon H^2 \in L_{\text{var}}([M, M]) \text{ and } \left(H^2 \overset{s}{\cdot} [M, M] \right)^{p/2} \in \mathscr{A}^+_{\text{loc}} \right\}.$$

It is clear that $L^p(M) \subseteq L^p_{\text{loc}}(M)$, $L^p(M) \subseteq L^r(M)$ and $L^p_{\text{loc}}(M) \subseteq L^r_{\text{loc}}(M)$, if $p \geqslant r \geqslant 1$. In view of the above remark, the class $L^2(M)$ (respectively $L^2_{\text{loc}}(M)$) has the previous meaning in the case $M \in \mathscr{M}^2$ (respectively $M \in \mathscr{M}^2_{\text{loc}}$).

Let H be a predictable locally bounded process. It follows from lemma 2.27 that $H \in L^1_{\text{loc}}(M)$ for any $M \in \mathscr{M}_{\text{loc}}$.

EXERCISE 3.4.– Prove that $L^p(M)$ and $L^p_{\text{loc}}(M)$ are linear spaces for every $p \geqslant 1$.

DEFINITION 3.3.– Let $M \in \mathscr{M}_{\text{loc}}$ and $H \in L^1_{\text{loc}}(M)$. The *stochastic integral* process $H \cdot M$ is the unique element in $\mathscr{M}_{\text{loc}}$ satisfying

$$H \cdot M_0 = 0, \quad (H \cdot M)^c = H \cdot M^c, \quad \Delta(H \cdot M) = H \Delta M.$$

In this definition, the process $H \cdot M^c$ is understood as in definition 3.2.

PROPOSITION 3.2.– Definition 3.3 is correct.

PROOF.– According to the definition of the quadratic variation,

$$H^2 \overset{s}{\cdot} [M, M] = H^2 \overset{s}{\cdot} \langle M^c, M^c \rangle + S(H^2(\Delta M)^2). \tag{3.5}$$

In particular, $H^2 \overset{s}{\cdot} \langle M^c, M^c \rangle \in \mathcal{V}^+$. The process $H^2 \overset{s}{\cdot} \langle M^c, M^c \rangle$ is a.s. continuous. Hence, $H^2 \overset{s}{\cdot} \langle M^c, M^c \rangle \in \mathscr{A}_{\text{loc}}^+$, i.e., $H \in L_{\text{loc}}^2(M^c)$. Therefore, the integral $H \cdot M^c$ is defined and is a continuous local martingale, which follows from $\Delta(H \cdot M^c) = H\Delta M^c = 0$.

Put $X := H\Delta M$. Let us check that X satisfies the assumptions of theorem 2.33. The relation [2.68] is obvious from the assumption $H \in L_{\text{loc}}^1(M)$ and [3.5]. Let T be a predictable stopping time. By theorem 2.10 and proposition 1.11, a.s.

$$\mathsf{E}(|X_T|\mathbb{1}_{\{T<\infty\}}|\mathscr{F}_{T-}) = |H_T|\mathbb{1}_{\{T<\infty\}}\mathsf{E}(|\Delta M_T|\mathbb{1}_{\{T<\infty\}}|\mathscr{F}_{T-}) < \infty$$

and

$$\mathsf{E}(X_T\mathbb{1}_{\{T<\infty\}}|\mathscr{F}_{T-}) = H_T\mathbb{1}_{\{T<\infty\}}\mathsf{E}(\Delta M_T\mathbb{1}_{\{T<\infty\}}|\mathscr{F}_{T-}) = 0.$$

Hence, [2.67] is also satisfied. We conclude from theorem 2.33 that there exists an $N \in \mathscr{M}_{\text{loc}}^d$ such that $\Delta N = H\Delta M$. Thus, we can take $H \cdot M^c + N$ as $H \cdot M$. The uniqueness is evident from theorem 2.31.   □

Definition 3.3 of the stochastic integral coincides with previous definition 3.2 in the case $M \in \mathscr{M}_{\text{loc}}^2$, $H \in L_{\text{loc}}^2(M)$. This follows from the above mentioned properties of the stochastic integral in the sense of definition 3.2.

We now turn to properties of the stochastic integral that we just defined. The first assertion is an immediate consequence of definition 3.3 and [3.5].

PROPOSITION 3.3.– Let $M \in \mathscr{M}_{\text{loc}}$ and $H \in L_{\text{loc}}^1(M)$. Then, $H \in L_{\text{loc}}^1(M^c) \cap L_{\text{loc}}^1(M^d)$, $(H \cdot M)^c = H \cdot M^c$, $(H \cdot M)^d = H \cdot M^d$.

The next proposition follows from the properties of the stochastic integral in the sense of definition 3.2, exercise 3.4 and definition 3.3.

PROPOSITION 3.4.– Let $M \in \mathscr{M}_{\text{loc}}$, $H^1, H^2 \in L_{\text{loc}}^1(M)$ and $\alpha, \beta \in \mathbb{R}$. Then, $\alpha H^1 + \beta H^2 \in L_{\text{loc}}^1(M)$ and $(\alpha H^1 + \beta H^2) \cdot M = \alpha(H^1 \cdot M) + \beta(H^2 \cdot N)$.

Next two propositions are proved similarly.

PROPOSITION 3.5.– Let $M, N \in \mathscr{M}_{\text{loc}}$, $H \in L_{\text{loc}}^1(M) \cap L_{\text{loc}}^1(N)$, $\alpha, \beta \in \mathbb{R}$. Then, $H \in L_{\text{loc}}^1(\alpha M + \beta N)$ and $H \cdot (\alpha M + \beta N) = \alpha(H \cdot M) + \beta(H \cdot N)$.

PROPOSITION 3.6.– Let $M \in \mathcal{M}_{\mathrm{loc}}$, $H \in L^1_{\mathrm{loc}}(M)$, and T be a stopping time. Then

$$(H \cdot M)^T = H \cdot M^T = (H \mathbb{1}_{\rrbracket 0, T \rrbracket}) \cdot M.$$

EXERCISE 3.5.– Prove propositions 3.5 and 3.6.

THEOREM 3.4.– Let $M \in \mathcal{M}_{\mathrm{loc}}$ and $H \in L^1_{\mathrm{loc}}(M)$.

1) For every $N \in \mathcal{M}_{\mathrm{loc}}$,

$$H \in L_{\mathrm{var}}([M, N]) \quad \text{and} \quad [H \cdot M, N] = H \overset{s}{\cdot} [M, N].$$

In particular,

$$[H \cdot M, H \cdot M] = H^2 \overset{s}{\cdot} [M, M]. \tag{3.6}$$

If $L \in \mathcal{M}_{\mathrm{loc}}$, $L_0 = 0$ and $[L, N] = H \overset{s}{\cdot} [M, N]$ for every $N \in \mathcal{M}_{\mathrm{loc}}$, then $L = H \cdot M$.

2) For $p \geqslant 1$,

$$H \cdot M \in \mathcal{H}^p \Leftrightarrow H \in L^p(M),$$
$$H \cdot M \in \mathcal{H}^p_{\mathrm{loc}} \Leftrightarrow H \in L^p_{\mathrm{loc}}(M).$$

PROOF.–

1) That $H \in L_{\mathrm{var}}([M, N])$ holds follows from the Kunita–Watanabe inequality [2.54] (we have mentioned in section 2.8 that it is also true for $M, N \in \mathcal{M}_{\mathrm{loc}}$). Next, due to properties of the stochastic integral in the sense of definition 3.2,

$$
\begin{aligned}
[H \cdot M, N] &= \langle (H \cdot M)^c, N^c \rangle + S(\Delta(H \cdot M)\Delta N) \\
&= \langle H \cdot M^c, N^c \rangle + S(H \Delta M \Delta N) \\
&= H \overset{s}{\cdot} \langle M^c, N^c \rangle + H \overset{s}{\cdot} S(\Delta M \Delta N) = H \overset{s}{\cdot} [M, N].
\end{aligned}
$$

The last assertion is immediate.

2) Both assertions follow immediately from corollary 2.18 and [3.6]. □

PROPOSITION 3.7.– Let $M \in \mathcal{M}_{\mathrm{loc}}$, $H \in L^1_{\mathrm{loc}}(M)$, K be a predictable process, $p \geqslant 1$. Then,

$$K \in L^p(H \cdot M) \Leftrightarrow KH \in L^p(M),$$
$$K \in L^p_{\mathrm{loc}}(H \cdot M) \Leftrightarrow KH \in L^p_{\mathrm{loc}}(M),$$

and any one of these relations implies $K \cdot (H \cdot M) = (KH) \cdot M$.

EXERCISE 3.6.– Prove proposition 3.7.

If a local martingale M is also a process with finite variation, then there are two classes of predictable integrands, $L_{\text{loc}}^1(M)$ and $L_{\text{var}}(M)$, to which correspond two kinds of integrals, $H \cdot M \in \mathscr{M}_{\text{loc}}$ and $H \overset{s}{\cdot} M \in \mathscr{V}$. The next theorem gives a full description of relationships between them. Recall that $\mathscr{M}_{\text{loc}} \cap \mathscr{V} \subseteq \mathscr{A}_{\text{loc}}$ (lemma 2.6), and also the fact that $H \in L_{\text{var}}(M)$ and $H \overset{s}{\cdot} M \in \mathscr{A}_{\text{loc}}$ imply $H \overset{s}{\cdot} M \in \mathscr{M}_{\text{loc}}$ (proposition 2.10 (4)).

THEOREM 3.5.– Let $M \in \mathscr{M}_{\text{loc}} \cap \mathscr{V}$.

1) If $H \in L_{\text{loc}}^1(M) \cap L_{\text{var}}(M)$, then $H \cdot M = H \overset{s}{\cdot} M$.

2) If $H \in L_{\text{loc}}^1(M) \setminus L_{\text{var}}(M)$, then $H \cdot M \notin \mathscr{V}$.

3) If $H \in L_{\text{var}}(M) \setminus L_{\text{loc}}^1(M)$, then $H \overset{s}{\cdot} M \notin \mathscr{M}_{\text{loc}}$.

The most typical situation is described, of course, in the first statement of the theorem. For example, if H is a locally bounded predictable process, then $H \in L_{\text{loc}}^1(M) \cap L_{\text{var}}(M)$ for every $M \in \mathscr{M}_{\text{loc}} \cap \mathscr{V}$. For an example of a local martingale $M \in \mathscr{M}_{\text{loc}} \cap \mathscr{V}$ and a process $H \in L_{\text{var}}(M)$ such that $H \overset{s}{\cdot} M \notin \mathscr{M}_{\text{loc}}$, see example 2.3; it follows from (1) that $H \notin L_{\text{loc}}^1(M)$ in this example. An example of a local martingale $M \in \mathscr{M}_{\text{loc}} \cap \mathscr{V}$ and a process $H \in L_{\text{loc}}^1(M) \setminus L_{\text{var}}(M)$ will be constructed after the proof of the theorem.

PROOF.–

1) Let $H \in L_{\text{loc}}^1(M) \cap L_{\text{var}}(M)$. By the definition of $L_{\text{loc}}^1(M)$, there is a localizing sequence $\{S_n\}$ such that:

$$\mathsf{E}(H^2 \overset{s}{\cdot} [M, M]_{S_n})^{1/2} < \infty$$

for each n. Put

$$T_n := S_n \wedge \inf\{t \colon \text{Var}\,(H \overset{s}{\cdot} M)_t \geqslant n\}.$$

Then $T_n \uparrow \infty$ a.s. due to the condition $H \in L_{\text{var}}(M)$, and

$$\text{Var}\,(H \overset{s}{\cdot} M)_{T_n} \leqslant n + |H_{T_n} \Delta M_{T_n} \mathbb{1}_{\{T_n < \infty\}}| \leqslant n + (H^2 \overset{s}{\cdot} [M, M]_{S_n})^{1/2}.$$

The right-hand side has a finite expectation, hence $H \overset{s}{\cdot} M \in \mathscr{A}_{\text{loc}}$.

Next, by proposition 2.10 (4), $H \overset{s}{\cdot} M \in \mathscr{M}_{\text{loc}}$ and, moreover, by lemma 2.24, $H \overset{s}{\cdot} M \in \mathscr{M}_{\text{loc}}^d$. Since $M \in \mathscr{M}_{\text{loc}}^d$ by lemma 2.24, we have $(H \cdot M)^c = H \cdot M^c = 0$.

Moreover, $\Delta(H \overset{s}{\cdot} M) = H\Delta M$. Therefore, $H \overset{s}{\cdot} M$ coincides with $H \cdot M$ according to the definition of the latter integral.

2) Assume that $H \in L^1_{\text{loc}}(M)$ and $H \cdot M \in \mathscr{V}$. Put $A := \sum_{0 < s \leqslant \cdot} \Delta M_s$ and recall that $A \in \mathscr{A}_{\text{loc}}$ and $M = A - \widetilde{A}$, where $\widetilde{A}$ is the compensator of A, see proposition 2.10 (5). Since $H \cdot M \in \mathscr{M}_{\text{loc}}$ by the definition of the integral and $H \cdot M \in \mathscr{V}$ by the assumption, we get $H \cdot M \in \mathscr{A}_{\text{loc}}$ by lemma 2.6. But we also have $\Delta(H \cdot M) = H\Delta M$ and, hence, the process $H \overset{s}{\cdot} A$ is defined. Moreover,

$$\text{Var}\,(H \overset{s}{\cdot} A) = \sum_{0 < s \leqslant \cdot} |H_s \Delta A_s| = \sum_{0 < s \leqslant \cdot} |H_s \Delta M_s| = \sum_{0 < s \leqslant \cdot} |\Delta(H \cdot M)_s|$$

$$\leqslant \text{Var}\,(H \cdot M),$$

where $H \overset{s}{\cdot} A \in \mathscr{A}_{\text{loc}}$. It follows from theorem 2.21 (2) that $H \in L_{\text{var}}(\widetilde{A})$. Thus, $H \in L_{\text{var}}(M)$.

3) Assume that $H \in L_{\text{var}}(M)$ and $H \overset{s}{\cdot} M \in \mathscr{M}_{\text{loc}}$. Then $H \overset{s}{\cdot} M \in \mathscr{A}_{\text{loc}}$ by lemma 2.6. As we noted above, $M \in \mathscr{M}^d_{\text{loc}}$, therefore, $[M, M] = S((\Delta M)^2)$ and

$$(H^2 \overset{s}{\cdot} [M, M])^{1/2} = \left\{ S((H\Delta M)^2) \right\}^{1/2} \leqslant S(|H\Delta M|)$$

$$\leqslant \text{Var}\,(H \overset{s}{\cdot} M) \in \mathscr{A}^+_{\text{loc}},$$

hence $H \in L^1_{\text{loc}}(M)$.   □

EXAMPLE 3.2.– Let $\xi_1, \ldots, \xi_n, \ldots$ be independent random variables on a complete probability space $(\Omega, \mathscr{F}, \mathsf{P})$. Moreover, $\mathsf{P}(\xi_n = \pm 2^{-n}) = 1/2$. Put $t_n = n/(n+1)$, $\mathscr{F}_t := \sigma\{\xi_1, \ldots, \xi_n, \ldots : t_n \leqslant t\} \vee \sigma\{\mathscr{N}\}$, where $\mathscr{N}$ consists of null sets from $\mathscr{F}$,

$$M_t := \sum_{n:\, t_n \leqslant t} \xi_n.$$

It is clear that $M = (M_t)$ is a martingale and

$$[M, M]_t = \sum_{n:\, t_n \leqslant t} 2^{-2n}, \quad \text{Var}\,(M)_t = \sum_{n:\, t_n \leqslant t} 2^{-n} \leqslant 1,$$

so that $M \in \mathscr{A}$. Let us define deterministic functions $H = (H_t)$ as follows: $H_{t_n} = 2^n/n$, $n = 1, 2, \ldots$ and $H_t = 0$, if $t \notin \{t_1, \ldots, t_n, \ldots\}$. Then

$$H^2 \overset{s}{\cdot} [M, M]_t = \sum_{n:\, t_n \leqslant t} \frac{2^{2n}}{n^2} 2^{-2n} = \sum_{n:\, t_n \leqslant t} n^{-2} \leqslant \frac{\pi^2}{6}.$$

It follows that $H \in L^p(M)$ for every $p \geqslant 1$. However,

$$\int\limits_0^1 |H_s| \, d \operatorname{Var}(M)_s = \sum_{n=1}^\infty \frac{2^n}{n} 2^{-n} = \sum_{n=1}^\infty \frac{1}{n} = \infty,$$

so that $H \notin L_{\operatorname{var}}(M)$.

3.2. Semimartingales. Stochastic integrals with respect to semimartingales: locally bounded integrands. Itô's formula

DEFINITION 3.4.– An adapted càdlàg process $X = (X_t)_{t \in \mathbb{R}_+}$ is called a *semimartingale* if X admits a decomposition:

$$X = X_0 + M + A, \quad M \in \mathscr{M}_{\mathrm{loc}}, \quad A \in \mathscr{V}. \tag{3.7}$$

The class of all semimartingales is denoted by $\mathscr{S}$.

Trivial examples of semimartingales are, of course, local martingales and processes with finite variation. More interesting examples of semimartingales are submartingales and supermartingales, see the Doob–Meyer decomposition (theorem 2.25).

The decomposition [3.7] is not unique. However, it follows from corollary 2.14 that if there is a decomposition [3.7] with a predictable A, then such a decomposition is unique (up to an evanescent set).

DEFINITION 3.5.– A semimartingale X is called a *special semimartingale* if there exists a decomposition [3.7] with a predictable A. This decomposition is called the *canonical decomposition* of a special semimartingale. The class of all special semimartingales is denoted by $\mathscr{S}_p$.

Local martingales, submartingales and supermartingales are special semimartingales. Theorem 2.21 (i) asserts that a process with finite variation is a special semimartingale if and only if it is a process of locally integrable variation.

Before we provide a characterization of special semimartingales, let us prove a lemma that will be useful later as well.

LEMMA 3.2.– For every semimartingale X and for every $t \in \mathbb{R}_+$,

$$\sum_{0 < s \leqslant t} (\Delta X_s)^2 < +\infty \quad \text{a.s.}$$

PROOF.– Let $X = X_0 + M + A$, $M \in \mathcal{M}_{\mathrm{loc}}$, $A \in \mathcal{V}$. By the Schwarz inequality,

$$\left(\sum_{0<s\leqslant t} (\Delta X_s)^2 \right)^{1/2} \leqslant \left(\sum_{0<s\leqslant t} (\Delta M_s)^2 \right)^{1/2} + \left(\sum_{0<s\leqslant t} (\Delta A_s)^2 \right)^{1/2}. \qquad [3.8]$$

It remains for us to note that, by lemma 2.25,

$$\sum_{0<s\leqslant t} (\Delta M_s)^2 < +\infty \quad \text{a.s.}$$

and

$$\sum_{0<s\leqslant t} (\Delta A_s)^2 \leqslant \left(\sum_{0<s\leqslant t} |\Delta A_s| \right)^2 \leqslant \left\{ \mathrm{Var}\,(A)_t \right\}^2 < +\infty. \qquad [3.9]$$

$\square$

THEOREM 3.6.– Let X be a semimartingale. The following statements are equivalent:

1) X is a special semimartingale;

2) there exists a decomposition [3.7] with $A \in \mathcal{A}_{\mathrm{loc}}$;

3) $A \in \mathcal{A}_{\mathrm{loc}}$ in every decomposition [3.7];

4) $(X - X_0)^* \in \mathcal{A}_{\mathrm{loc}}^+$;

5) $\left\{ S((\Delta X)^2) \right\}^{1/2} \in \mathcal{A}_{\mathrm{loc}}^+$;

6) $(\Delta X)^* \in \mathcal{A}_{\mathrm{loc}}^+$.

PROOF.– $(1) \Rightarrow (2)$ follows from lemma 2.5.

$(2) \Rightarrow (1)$ Let X admit a decomposition [3.7] with $A \in \mathcal{A}_{\mathrm{loc}}$. By theorem 2.21 (1), there exists the compensator $\widetilde{A}$ of A. Then the decomposition $X = X_0 + (M + A - \widetilde{A}) + \widetilde{A}$ is a decomposition of the form [3.7] with a predictable process with finite variation.

$(2) \Rightarrow (4)$ Let X admit a decomposition [3.7] with $A \in \mathcal{A}_{\mathrm{loc}}$. Then, $(X - X_0)^* \leqslant M^* + A^* \leqslant M^* + \mathrm{Var}\,(A)$. By lemma 2.7, $M^* \in \mathcal{A}_{\mathrm{loc}}^+$, and (4) follows.

$(4) \Rightarrow (6)$ follows from the inequality $(\Delta X)^* \leqslant 2(X - X_0)^*$.

(6)$\Rightarrow$(3) Let us take any decomposition [3.7]. Since $(\Delta A)^* \leqslant (\Delta M)^* + (\Delta X)^*$ and $(\Delta M)^* \in \mathscr{A}_{\mathrm{loc}}^+$ by lemma 2.7, we have $(\Delta A)^* \in \mathscr{A}_{\mathrm{loc}}^+$. Hence, there exists a localizing sequence $\{S_n\}$ such that:

$$\mathsf{E} \sup_{s \leqslant S_n} |\Delta A_s| < \infty$$

for every n. Put

$$T_n := S_n \wedge \inf \{t \colon \mathrm{Var}\,(A) \geqslant n\}.$$

Then $T_n \uparrow \infty$ a.s. and

$$\mathrm{Var}\,(A)_{T_n} \leqslant n + |\Delta A_{T_n} \mathbb{1}_{\{T_n < \infty\}}| \leqslant n + \sup_{s \leqslant S_n} |\Delta A_s|,$$

hence, $\mathsf{E}\,\mathrm{Var}\,(A)_{T_n} < \infty$.

(3)$\Rightarrow$(2) is obvious.

(2)$\Rightarrow$(5) Let X admit a decomposition [3.7] with $A \in \mathscr{A}_{\mathrm{loc}}$. Then, by the inequalities [3.8] and [3.9],

$$\left\{ S((\Delta X)^2) \right\}^{1/2} \leqslant \left\{ S((\Delta M)^2) \right\}^{1/2} + \mathrm{Var}\,(A).$$

The first term on the right belongs to $\mathscr{A}_{\mathrm{loc}}^+$ by lemma 2.27, and the second term does the same by the assumption.

(5)$\Rightarrow$(6) follows from the inequality $(\Delta X)^* \leqslant \left\{ S((\Delta X)^2) \right\}^{1/2}$.   □

COROLLARY 3.8.– The following statements are equivalent:

1) X is a predictable semimartingale;

2) X is a special semimartingale and the local martingale M in its canonical decomposition is a.s. continuous.

In particular, a continuous semimartingale X is a special semimartingale and we can take continuous versions of M and A in its canonical decomposition.

PROOF.– Implication (2)$\Rightarrow$(1) is obvious due to corollary 1.2. Let X be a predictable semimartingale. By lemma 3.2, the process $S((\Delta X)^2)$ is well defined, and it is predictable and increasing by lemma 2.19. Then $\left\{ S((\Delta X)^2) \right\}^{1/2}$ is also a predictable increasing process. By lemma 2.5, $\left\{ S((\Delta X)^2) \right\}^{1/2} \in \mathscr{A}_{\mathrm{loc}}^+$, and it follows from theorem 3.6 that X is a special semimartingale. Let $X = X_0 + M + A$ be its canonical decomposition. Then M is a predictable local martingale. Combining proposition 1.11 and theorem 2.10, we obtain that $\Delta M_T \mathbb{1}_{\{T < \infty\}} = 0$ a.s. for every predictable stopping time T. By theorem 1.18, the set $\{\Delta M \neq 0\}$ is evanescent.

If X is continuous, then $\Delta M = -\Delta A$ and, as was proved above, A is a.s. continuous. So the decomposition

$$X = X_0 + \left(M + \sum_{s \leqslant \cdot} \Delta A_s \right) + \left(A - \sum_{s \leqslant \cdot} \Delta A_s \right)$$

is a required one. $\square$

The last assertion in corollary 3.8 follows also from the next theorem..

THEOREM 3.7.– Let X be a semimartingale and $|\Delta X| \leqslant a$ for some $a \in \mathbb{R}_+$. Then X is a special semimartingale and, for M and A in its canonical decomposition, $|\Delta A| \leqslant a$ and $|\Delta M| \leqslant 2a$.

PROOF.– By theorem 3.6, X is a special semimartingale. Let $X = X_0 + M + A$ be its canonical decomposition, and let T be a predictable stopping time. By proposition 1.11 and theorem 2.10,

$$\Delta A_T \mathbb{1}_{\{T < \infty\}} = \mathsf{E}\big(\Delta A_T \mathbb{1}_{\{T < \infty\}} \big| \mathscr{F}_{T-} \big) = \mathsf{E}\big(\Delta X_T \mathbb{1}_{\{T < \infty\}} \big| \mathscr{F}_{T-} \big)$$
$$- \mathsf{E}\big(\Delta M_T \mathbb{1}_{\{T < \infty\}} \big| \mathscr{F}_{T-} \big) = \mathsf{E}\big(\Delta X_T \mathbb{1}_{\{T < \infty\}} \big| \mathscr{F}_{T-} \big),$$

hence $|\Delta A_T \mathbb{1}_{\{T < \infty\}}| \leqslant a$ a.s. By theorem 1.18, the set $|\Delta A| > a$ is evanescent. $\square$

It is clear that the classes $\mathscr{S}$ and $\mathscr{S}_p$ are stable under stopping and linear operations. Let us show that they are stable under localization.

THEOREM 3.8.– Let X be an adapted càdlàg process. Assume that there is a localizing sequence $\{T_n\}$ of stopping times such that X^{T_n} is a semimartingale (respectively a special semimartingale) for every n. Then X is a semimartingale (respectively a special semimartingale).

PROOF.– Assume first that all X^{T_n} are semimartingales. By the definition, for every n, there are $M^n \in \mathscr{M}_{\mathrm{loc}}$ and $A^n \in \mathscr{V}$ such that

$$X^{T_n} = X_0 + M^n + A^n.$$

Define processes M and A as the results of "gluing" from $\{T_n\}$, $\{M^n\}$ and $\{T_n\}$, $\{A^n\}$, respectively; then $M \in \mathscr{M}_{\mathrm{loc}}$ and $A \in \mathscr{V}$. It follows from the definition of "gluing" that $X_0 + M + A$ is a result of "gluing" from $\{T_n\}$, $\{X^{T_n}\}$. Hence, $X_0 + M^{T_n} + A^{T_n}$ and X^{T_n} are indistinguishable for every n, i.e., $X_0 + M + A$ and X are indistinguishable.

The case of special semimartingales may be considered similarly (and even simpler since we can use the uniqueness of the canonical decomposition). An alternative proof is to use that X is a semimartingale, as we have just proved. So, it is enough to check any of statements (4)–(6) in theorem 3.6, which follow from $\mathscr{A}_{\text{loc}}^+ = (\mathscr{A}_{\text{loc}}^+)_{\text{loc}}$. $\square$

Let X be a semimartingale. Let us take two decompositions X of form [3.7]: $X = X_0 + M + A = X_0 + M' + A'$, $M, M' \in \mathscr{M}_{\text{loc}}$, $A, A' \in \mathscr{V}$. Then $M - M' = A' - A \in \mathscr{M}_{\text{loc}} \cap \mathscr{V}$, hence $M - M' \in \mathscr{M}_{\text{loc}}^d$ by lemma 2.24, hence $(M - M')^c = 0$, i.e., $M^c = M'^c$. Thus, the process M^c, where M is taken from a decomposition [3.7], is the same (up to an evanescent set) for any choice of a decomposition. This makes the next definition correct.

DEFINITION 3.6.– A continuous local martingale, denoted by X^c, is called the *continuous martingale component* of a semimartingale X if $X^c = M^c$ for every M satisfying [3.7].

The continuous martingale component X^c should not be confused with the continuous process $X - \sum_{0 < s \leqslant \cdot} \Delta X_s$, which is well defined, e.g., if $X \in \mathscr{V}$.

EXERCISE 3.7.– Show that, if X is a continuous semimartingale and $X^c = 0$, then $X - X_0$ is indistinguishable with a process with finite variation.

DEFINITION 3.7.– The *quadratic variation* of a semimartingale X is the increasing process defined by

$$[X, X] := \langle X^c, X^c \rangle + S\big((\Delta X)^2\big).$$

The *quadratic covariation* of semimartingales X and Y is the process with finite variation defined by

$$[X, Y] := \langle X^c, Y^c \rangle + S(\Delta X \Delta Y).$$

The correctness of this definition follows from lemma 3.2.

It is clear that, for local martingales, this definition coincides with definition 2.25. Let us also note that, if M is a local martingale and $[M, M] = 0$, then $M - M_0 = 0$, while, for a semimartingale X, $[X, X] = 0$ means that $X - X_0$ is indistinguishable with a continuous process with finite variation.

As in the case of local martingales, $[X, Y]$ is symmetric and bilinear, $[X, Y^T] = [X, Y]^T$ for every stopping time T and every $X, Y \in \mathscr{S}$. Also, the Kunita–Watanabe inequalities [2.54] and [2.56] remain true.

We now turn to the definition of a stochastic integral with respect to semimartingales. The definition in the general case will be given in section 3.4, and here we consider the case when an integrand H is a locally bounded predictable process, which is sufficient to formulate Itô's formula. In this section, there are integrals of three different types: the pathwise Lebesgue–Stieltjes integral defined in theorem 2.11, the stochastic integral with respect to local martingales, see definition 3.3, and the stochastic integral with respect to semimartingales. But in all cases and for all types of integrals, we will deal with locally bounded predictable integrands in this section. Recall that, given a locally bounded predictable H, we have $H \in L_{\text{var}}(X)$ for every $X \in \mathscr{V}$, $H \in L^1_{\text{loc}}(X)$ for every $X \in \mathscr{M}_{\text{loc}}$, and in the case $X \in \mathscr{V} \cap \mathscr{M}_{\text{loc}}$ the pathwise Lebesgue–Stieltjes integral $H \overset{s}{\cdot} X$, and the stochastic integral with respect to a local martingale $H \cdot X$ coincide (theorem 3.5 (1)). This is the reason why the integrals of the first and second type are denoted by the same symbol in this section: $H \cdot X$. The same notation will be used in this section for integrals of the third type, i.e. stochastic integrals with respect to semimartingales. The reason is that it will follow directly from the definition that for locally bounded predictable H, the integrals of the first and the third type are the same if $X \in \mathscr{V}$, and the integrals of the second and third types coincide if $X \in \mathscr{M}_{\text{loc}}$.

The idea how to define the stochastic integral with respect to a semimartingale is simple: we define $H \cdot X$ as $H \cdot M + H \cdot A$, where M and A are taken from a decomposition [3.7]. The next lemma shows that this definition does not depend on the choice of a decomposition. Let us note that $H \cdot M \in \mathscr{M}_{\text{loc}}$ and $H \cdot A \in \mathscr{V}$, so that $H \cdot M + H \cdot A \in \mathscr{S}$.

LEMMA 3.3.– Let X be a semimartingale and H be a locally bounded predictable process,

$$X = X_0 + M + A = X_0 + M' + A', \quad M, M' \in \mathscr{M}_{\text{loc}}, \quad A, A' \in \mathscr{V}.$$

Then

$$H \cdot M + H \cdot A = H \cdot M' + H \cdot A'.$$

PROOF.– Since $M - M' = A' - A \in \mathscr{M}_{\text{loc}} \cap \mathscr{V}$,

$$H \cdot (M - M') = H \cdot (A' - A)$$

by theorem 3.5 (1).   □

DEFINITION 3.8.– Let X be a semimartingale and H be a locally bounded predictable process. The *stochastic integral* of H with respect to X is a semimartingale, denoted by $H \cdot X$, such that, up to an evanescent set,

$$H \cdot X = H \cdot M + H \cdot A$$

for every M and A satisfying [3.7].

In the following propositions, we study the basic properties of the stochastic integral that has just been defined.

PROPOSITION 3.8.– Let X be a special semimartingale and H be a locally bounded predictable process. Then $H \cdot X$ is a special semimartingale with the canonical decomposition:

$$H \cdot X = H \cdot M + H \cdot A,$$

where $X = X_0 + M + A$ is the canonical decomposition of X.

PROOF.– We have $H \cdot M \in \mathscr{M}_{\mathrm{loc}}$, $H \cdot A \in \mathscr{V}$, and the process $H \cdot A$ is predictable by theorem 2.11.   □

PROPOSITION 3.9.– Let $X, Y \in \mathscr{S}$, and let H and K be locally bounded predictable processes, $\alpha, \beta \in \mathbb{R}$. Then:

$$H \cdot (\alpha X + \beta Y) = \alpha(H \cdot X) + \beta(H \cdot Y),$$
$$(\alpha H + \beta K) \cdot X = \alpha(H \cdot X) + \beta(K \cdot X).$$

PROOF.– The assertion follows from definition 3.8 and the corresponding properties of the stochastic integral with respect to local martingales (propositions 3.4 and 3.5) and the Lebesgue–Stieltjes integral (theorem 2.12).   □

PROPOSITION 3.10.– Let X be a semimartingale and H be a locally bounded predictable process. Then $(H \cdot X)^c = H \cdot X^c$.

PROOF.– The assertion follows from definitions 3.8 and 3.6 and proposition 3.3.   □

PROPOSITION 3.11.– Let X be a semimartingale, H be a locally bounded predictable process and T be a stopping time. Then

$$(H \cdot X)^T = H \cdot X^T = (H \mathbb{1}_{]\hspace{-0.5pt}]0,T]\hspace{-0.5pt}]}) \cdot X.$$

PROOF.– The assertion follows from definition 3.8 and the corresponding properties of the stochastic integral with respect to local martingales (propositions 3.6) and the Lebesgue–Stieltjes integral (theorem 2.12).   □

PROPOSITION 3.12.– Let X be a semimartingale and H be a locally bounded predictable process. Then

$$\Delta(H \cdot X) = H \Delta X.$$

PROOF.– The assertion follows from definition 3.8 and the corresponding properties of the stochastic integral with respect to local martingales (definition 3.3) and the Lebesgue–Stieltjes integral (theorem 2.12).   □

PROPOSITION 3.13.– Let X be a semimartingale and H be a locally bounded predictable process. Then for every semimartingale Y:

$$[H \cdot X, Y] = H \cdot [X, Y].$$

PROPOSITION 3.14.– Let X be a semimartingale, H and K be two locally bounded predictable processes. Then, $K \cdot (H \cdot X) = (KH) \cdot X$.

EXERCISE 3.8.– Prove propositions 3.13 and 3.14.

THEOREM 3.9.– Let X be a semimartingale and $\{H^n\}$ a sequence of predictable processes, which converges as $n \to \infty$ for all t and ω to a process H. Assume also that $|H^n| \leqslant K$ for all n, where K is a locally bounded predictable process. Then $\sup_{s \leqslant t} |H^n \cdot X_s - H \cdot X_s| \xrightarrow{P} 0$ as $n \to \infty$ for every $t \in \mathbb{R}_+$.

PROOF.– It is enough to consider two cases separately: $X = M \in \mathcal{M}_{\mathrm{loc}}$ and $X = A \in \mathcal{V}$. In the second case, the proof is simple: for almost all ω,

$$\sup_{s \leqslant t} |H^n \cdot A_s - H \cdot A_s| \leqslant \int_0^t |H_s^n - H_s| \, d \operatorname{Var}(A)_s \to 0$$

by the Lebesgue dominated convergence theorem, because a trajectory $K.(\omega)$ is bounded on $]0, t]$ for almost all ω (by a constant depending on ω).

Let now $M \in \mathcal{M}_{\mathrm{loc}}$ be given. We take a localizing sequence $\{T_n\}$ such that $\mathsf{E}[M, M]_{T_n}^{1/2} < \infty$ and $|K| \mathbb{1}_{]0, T_n]} \leqslant C_n < \infty$. By Davis' inequality and theorem 3.4,

$$\mathsf{E} \sup_{s \leqslant T_n} |H^n \cdot M_s - H \cdot M_s| \leqslant C \mathsf{E}[(H^n - H) \cdot M, (H^n - H) \cdot M]_{T_n}^{1/2}$$

$$= C \mathsf{E} \big((H^n - H)^2 \cdot [M, M]_{T_n} \big)^{1/2}.$$

Applying twice the dominated convergence theorem, we first obtain $(H^n - H)^2 \cdot [M, M]_{T_n} \to 0$ a.s., and, second, that the right-hand side of the previous formula tends to 0. In particular,

$$\sup_{s \leqslant T_n} |H^n \cdot M_s - H \cdot M_s| \xrightarrow{P} 0.$$

It remains to note that:

$$P\left(\sup_{s\leqslant t}|H^n\cdot M_s - H\cdot M_s| > \varepsilon\right)$$

$$\leqslant P\left(\sup_{s\leqslant T_n}|H^n\cdot M_s - H\cdot M_s| > \varepsilon\right) + P(T_n < t).\qquad\square$$

We now turn to the most fundamental result of stochastic calculus, Itô's formula, which is given without the proof. We say that a process $X = (X^1,\ldots,X^d)$ with values in $\mathbb{R}^d$ is a d-dimensional semimartingale if all components $X^1,\ldots,X^d$ are semimartingales.

THEOREM 3.10 (Itô's formula).– Let $X = (X^1,\ldots,X^d)$ be a d-dimensional semimartingale, and let F be a twice continuously differentiable function from $\mathbb{R}^d$ to $\mathbb{R}$. Then $F(X)$ is a semimartingale and

$$F(X) = F(X_0) + \sum_{i=1}^d \frac{\partial F}{\partial x_i}(X_-)\cdot X^i + \frac{1}{2}\sum_{i,j=1}^d \frac{\partial^2 F}{\partial x_i\partial x_j}(X_-)\cdot\langle(X^i)^c,(X^j)^c\rangle$$

$$+ S\left(F(X) - F(X_-) - \sum_{i=1}^d \frac{\partial F}{\partial x_i}(X_-)\Delta X^i\right).$$

Let us discuss the contents of Itô's formula in detail. Equality of the left and right sides is understood, of course, up to an evanescent set. On the left side, we have a real-valued adapted process $F(X)$. Since F is continuous, its paths are càdlàg. On the right side, we have the sum of four terms. The first term is simply the value of the process $F(X)$ at time 0. Further, for each i, $Y^i := \frac{\partial F}{\partial x_i}(X)$ is an adapted càdlàg process (for the same reasons as before). Therefore, $Y_-^i = \frac{\partial F}{\partial x_i}(X_-)$ is a predictable locally bounded (and left-continuous) process, which is integrated with respect to the semimartingale X^i in the second term. In particular, by proposition 3.8, the second term on the right is a special semimartingale if these are all X^i. Similarly, $\frac{\partial^2 F}{\partial x_i\partial x_j}(X_-)$ is a predictable locally bounded (and left-continuous) process, which is integrated with respect to the continuous process with finite variation $\langle(X^i)^c,(X^j)^c\rangle$. Therefore, the third term is a continuous process with finite variation. Let us also note that here we deal with the Lebesgue–Stieltjes integral, and the predictability of the integrand does not play any role; in particular, the process X_- in the integrand can be replaced by X without changing the value of the integral because we integrate with respect to a continuous process. It remains to analyze the fourth term on the right. First, we show that it is well defined. Let $t\in\mathbb{R}_+$. Consider the trajectory $X.(\omega)$ on the interval $[0,t]$

for every ω. Since it is càdlàg, it does not come out of some compact $K = K(\omega, t)$. Denote:

$$C(\omega, t) := \sup_{x \in K(\omega,t)} \sup_{i,j} \left| \frac{\partial^2 F}{\partial x_i \partial x_j}(x) \right|.$$

It follows from Taylor's formula for functions of several variables that, for $s \leqslant t$,

$$\left| F(X_s(\omega)) - F(X_{s-}(\omega)) - \sum_{i=1}^{d} \frac{\partial F}{\partial x_i}(X_{s-}(\omega)) \Delta X_s^i(\omega) \right|$$

$$\leqslant \frac{C(\omega, t)}{2} \sum_{i,j=1}^{d} |\Delta X_s^i(\omega)| |\Delta X_s^j(\omega)| \leqslant \frac{C(\omega, t) d}{2} \sum_{i=1}^{d} (\Delta X_s^i(\omega))^2.$$

Therefore, by lemma 3.2, the fourth term is well defined and is a process with finite variation. Note that its jumps are exactly the same as you need in order to jumps in the left and right sides of Itô's formula match.

Since the right side of Itô's formula is a semimartingale, the left side is a semimartingale, too, as the theorem asserts. Thus, the class of semimartingales is closed with respect to a wide class of transformations. Note also that if all X^i are special semimartingales, then $F(X)$ is a special semimartingale if and only if the fourth term on the right is.

It often happens that there is a need to apply Itô's formula for the function F which is twice continuously differentiable (and perhaps even defined) only on an open subset of $\mathbb{R}^d$, while the process X takes values in this subset and, maybe, in its boundary. We show how this can be done in a special case. Let dimension $d = 1$, so that $X \in \mathscr{S}$ and takes nonnegative values, $X_0 > 0$, and the function F is defined and twice continuously differentiable on $(0, \infty)$. For each natural n, denote by F_n some function with values in $\mathbb{R}$, which is defined and twice continuously differentiable on $\mathbb{R}$ and coincides with F on $[1/n, \infty[$. We also put $T_n := \inf \{t \colon X_t < 1/n\}$. Applying Itô's formula to F_n and X, we obtain

$$F_n(X) = F_n(X_0) + F_n'(X_-) \cdot X + \frac{1}{2} F_n''(X_-) \cdot \langle X^c, X^c \rangle$$

$$+ S\big(F_n(X) - F_n(X_-) - F_n'(X_-) \Delta X\big).$$

Let us now stop the processes in both sides of this formula at T_n, and use proposition 3.11:

$$F_n(X)^{T_n} = F_n(X_0) + \{F_n'(X_-)\mathbb{1}_{]0,T_n]}\} \cdot X + \frac{1}{2}\{F_n''(X_-)\mathbb{1}_{]0,T_n]}\} \cdot \langle X^c, X^c \rangle$$
$$+ S\Big(\{F_n(X) - F_n(X_-) - F_n'(X_-)\Delta X\}\mathbb{1}_{]0,T_n]}\Big).$$

Note that $X \geqslant 1/n$ on $[0, T_n[$ and $X_- \geqslant 1/n$ on $]0, T_n]$, hence, $F_n(X_-)\mathbb{1}_{]0,T_n]} = F(X_-)\mathbb{1}_{]0,T_n]}$, $F_n'(X_-)\mathbb{1}_{]0,T_n]} = F'(X_-)\mathbb{1}_{]0,T_n]}$, $F_n''(X_-)\mathbb{1}_{]0,T_n]} = F''(X_-)\mathbb{1}_{]0,T_n]}$, and we can rewrite the formula as

$$F_n(X)^{T_n} = F_n(X_0) + \{F'(X_-)\mathbb{1}_{]0,T_n]}\} \cdot X + \frac{1}{2}\{F''(X_-)\mathbb{1}_{]0,T_n]}\} \cdot \langle X^c, X^c \rangle$$
$$+ S\Big(\{F_n(X) - F(X_-) - F'(X_-)\Delta X\}\mathbb{1}_{]0,T_n]}\Big).$$

Moreover, assume now that either $X > 0$ everywhere or F is defined at 0 (the right-continuity of F at 0 is not required here). Then:

$$F(X)^{T_n} = F(X_0) + \{F'(X_-)\mathbb{1}_{]0,T_n]}\} \cdot X + \frac{1}{2}\{F''(X_-)\mathbb{1}_{]0,T_n]}\} \cdot \langle X^c, X^c \rangle$$
$$+ S\Big(\{F(X) - F(X_-) - F'(X_-)\Delta X\}\mathbb{1}_{]0,T_n]}\Big), \qquad [3.10]$$

which can be easily seen comparing this formula with the previous formula.

If $X > 0$ and $X_- > 0$ everywhere, then $T_n \to \infty$, processes $F'(X_-)$ and $F''(X_-)$ are locally founded, and we arrive at Itô's formula in the standard form:

$$F(X) = F(X_0) + F'(X_-) \cdot X + \frac{1}{2}F''(X_-) \cdot \langle X^c, X^c \rangle$$
$$+ S\big(F(X) - F(X_-) - F'(X_-)\Delta X\big). \qquad [3.11]$$

If we impose only additional assumptions made before [3.10], then we can say, by abuse of language, that Itô's formula [3.11] holds on the set $\bigcup_n [0, T_n]$. Essentially, this means that [3.10] holds for every n.

EXERCISE 3.9.– Show that the process M^n in example 2.2 is a local martingale.

Itô's formula can be also generalized in some other directions. For example, Tanaka–Meyer formula is a generalization of Itô's formula for *convex* $F \colon \mathbb{R} \to \mathbb{R}$.

Itô's formula can be derived from its special case, the formula of integration by parts [3.12]. We also deduce two versions of the formula of integration by parts, [3.13] and [3.14], which are valid under additional assumptions.

THEOREM 3.11 (integration by parts).– Let X and Y be semimartingales. Then XY is a semimartingale and

$$XY = X_0 Y_0 + Y_- \cdot X + X_- \cdot Y + [X, Y].$$ [3.12]

If $Y \in \mathcal{V}$, then $[X, Y] = (\Delta X) \overset{s}{\cdot} Y$ and

$$XY = Y_- \cdot X + X \overset{s}{\cdot} Y.$$ [3.13]

If $Y \in \mathcal{V}$ is predictable, then $[X, Y] = (\Delta Y) \cdot X$ and

$$XY = Y \cdot X + X_- \cdot Y.$$ [3.14]

PROOF.– Applying Itô's formula with $F(x, y) = xy$, we obtain that XY is a semimartingale and

$$\begin{aligned} XY &= X_0 Y_0 + Y_- \cdot X + X_- \cdot Y + \langle X^c, Y^c \rangle \\ &\quad + S(XY - X_- Y_- - Y_- \Delta X - X_- \Delta Y) \\ &= X_0 Y_0 + Y_- \cdot X + X_- \cdot Y + \langle X^c, Y^c \rangle + S(\Delta X \Delta Y), \end{aligned}$$

i.e. [3.12] holds.

If $Y \in \mathcal{V}$, then $Y^c = 0$, hence,

$$[X, Y] = S(\Delta X \Delta Y) = (\Delta X) \overset{s}{\cdot} Y.$$

Now let $Y \in \mathcal{V}$ be predictable. Note that, in this case, the process $\mathrm{Var}\,(Y)$ is locally bounded, see the proof of lemma 2.5. Thus, Y is locally bounded and the integrals $(\Delta Y) \cdot X$ and $Y \cdot X$ are well defined.

Consider the semimartingale $Z = [X, Y] - (\Delta Y) \cdot X$. It is clear that $Z_0 = 0$ and Z is continuous. By corollary 3.8, Z is a special semimartingale, and the processes $M \in \mathcal{M}_{\mathrm{loc}}$ and $A \in \mathcal{V}$ in its canonical decomposition $Z = M + A$ are continuous. Further,

$$M = Z^c = [X, Y]^c - \big((\Delta Y) \cdot X \big)^c = -(\Delta Y) \cdot X^c,$$

hence

$$\langle M, M \rangle = (\Delta Y)^2 \cdot \langle X^c, X^c \rangle = 0,$$

because $\langle X^c, X^c \rangle$ is continuous, the integrand $(\Delta Y)^2$, for every ω, does not equal to zero in at most countable set, and we deal with the pathwise Lebesgue–Stieltjes integral. Hence, $M = 0$ and Z is a continuous process with finite variation.

Put $H := \mathbb{1}_{\{\Delta Y=0\}}$. Then H is a bounded predictable process and the sets $\{t: \Delta H_t(\omega) \neq 1\}$ are at most countable (for every ω). Therefore, the Lebesgue–Stieltjes integral $H \cdot Z$ is indistinguishable with Z. However, $Y^c = 0$, hence $[X, Y] = S(\Delta X \Delta Y)$ and

$$H \cdot [X, Y] = S(H \Delta X \Delta Y) = 0.$$

By proposition 3.14,

$$H \cdot ((\Delta Y) \cdot X) = (H \Delta Y) \cdot X = 0.$$

Thus, $Z = 0$.   □

COROLLARY 3.9.– If $M \in \mathscr{M}_{\mathrm{loc}}$, $A \in \mathscr{V}$, and A is predictable, then $A \cdot M \in \mathscr{M}_{\mathrm{loc}}$ and

$$AM = A \cdot M + M_- \cdot A$$

is the canonical decomposition of the special semimartingale AM.

EXERCISE 3.10.– Deduce Itô's formula for an arbitrary polynomial $F(x_1, \ldots, x_n)$ from the formula [3.12] of integration by parts.

3.3. Stochastic exponential

Let us consider the equation:

$$Z = 1 + Z_- \cdot X \tag{3.15}$$

with a *given* semimartingale X and an *unknown* process Z, which is assumed to belong to the class of adapted càdlàg processes (which guarantees that the stochastic integral is well defined). Equation [3.15] is usually written in the symbolic form

$$dZ = Z_- \, dX, \quad Z_0 = 1,$$

and is said to be a stochastic differential equation, though, essentially, [3.15] is a stochastic integral equation. Similarly to the ordinary differential equation $dz/dx = z$, its solution is called the *stochastic exponential* (or *Doléans exponential*) of the semimartingale X.

Before we state and prove the main result of this section, we recall how the infinite product is defined. If $p_1, p_2, \ldots, p_n, \ldots$ is a sequence of real numbers, then the symbol

$$p_1 \cdot p_2 \cdots p_n \cdots = \prod_{n=1}^{\infty} p_n$$

is called an infinite product. If there exists

$$\lim_{n \to \infty} \prod_{k=1}^{n} p_k,$$

then this limit is called the value of the infinite product. Moreover, if the limit is finite and *is different from* 0, then we say that the infinite product converges. It is necessary for the convergence of the infinite product that $p_n \neq 0$ for all n and $\lim_n p_n = 1$. Since discarding a finite number of (nonzero) terms in the product does not affect the convergence, it is sufficient to consider the question of the convergence of infinite products when all $p_n > 0$. Under this condition, the infinite product converges if and only if the infinite series $\sum_{n=1}^{\infty} \log p_n$ converges, and then

$$\prod_{n=1}^{\infty} p_n = \exp\left(\sum_{n=1}^{\infty} \log p_n \right).$$

If, moreover, the series $\sum_{n=1}^{\infty} \log p_n$ converges absolutely, then we say that the product $\prod_{n=1}^{\infty} p_n$ converges absolutely. If a product converges absolutely, then all its rearrangements converge and to the same limit.

It is more convenient for us to change this terminology and to use the following definition: an infinite product $\prod_{n=1}^{\infty} p_n$ is called convergent absolutely if $p_n \leqslant 0$ only for a finite number of n and the product $\prod_{n:\, p_n > 0} p_n$ converges absolutely in the previous sense. It is clear that this definition guarantees that the value of the product $\prod_{n=1}^{\infty} p_n$ is defined and finite and does not change under any rearrangement of factors. However, it may be equal to 0, which happens if and only if one of factors is 0.

LEMMA 3.4.– Let X be a semimartingale. Then, for almost all ω, the infinite product

$$\prod_{s \leqslant t} (1 + \Delta X_s) e^{-\Delta X_s}$$

converges absolutely for all $t \in \mathbb{R}_+$. There exists a process V such that $V_0 = 1$, $V - 1 \in \mathcal{V}^d$, and, for every $t \in \mathbb{R}_+$,

$$V_t = \prod_{s \leqslant t}(1 + \Delta X_s)e^{-\Delta X_s} \quad \text{a.s.} \tag{3.16}$$

If the set $\{\Delta X = -1\}$ is evanescent, there is a version of V which does not vanish. If $\Delta X \geqslant -1$, then we can choose a version of V, whose all trajectories are nonincreasing.

PROOF.– Since the set $\{s \leqslant t : |\Delta X_s(\omega)| > 1/2\}$ is finite for all ω and $t \in \mathbb{R}$, the process

$$V_t' = \prod_{s \leqslant t}(1 + \Delta X_s \mathbb{1}_{\{|\Delta X_s| > 1/2\}})e^{-\Delta X_s \mathbb{1}_{\{|\Delta X_s| > 1/2\}}}$$

is well defined and adapted, $V_0' = 1$, with all trajectories being piecewise constant (with a finite number of pieces on compact intervals) and right-continuous. Hence, $V' - 1 \in \mathcal{V}^d$. For a given t, the absolute convergence of the product

$$\prod_{s \leqslant t : |\Delta X_s| \leqslant 1/2} (1 + \Delta X_s)e^{-\Delta X_s}$$

is equivalent to the absolute convergence of the series

$$\sum_{s \leqslant t : |\Delta X_s| \leqslant 1/2} \{\log(1 + \Delta X_s) - \Delta X_s\},$$

which holds, in view of the inequality $-x^2 \leqslant \log(1 + x) - x \leqslant 0$, $-1/2 \leqslant x \leqslant 1/2$, for ω such that the series $\sum_{s \leqslant t}(\Delta X_s)^2$ converges, i.e. for almost all ω by lemma 3.2. Therefore, we can define the process

$$B := S\{\log(1 + \Delta X_s \mathbb{1}_{\{|\Delta X_s(\omega)| \leqslant 1/2\}}) - \Delta X_s \mathbb{1}_{\{|\Delta X_s(\omega)| \leqslant 1/2\}}\} \in \mathcal{V}^d;$$

moreover, we can assume that $-B \in \mathcal{V}^+$. Now put

$$V = V'e^B.$$

It is clear that [3.16] holds. Note that, by Itô's formula,

$$e^B - 1 = e^{B-} \cdot B + S(e^B - e^{B-} - e^{B-}\Delta B) = S(e^B - e^{B-}) \in \mathcal{V}^d,$$

where we have used that $e^{B-} \cdot B = S(e^{B-} \Delta B)$ due to $B \in \mathscr{V}^d$. Similarly, applying the formula of integration by parts, we get $V'e^B - 1 \in \mathscr{V}^d$. The last two assertions are clear from the construction. □

Also, we prove the uniqueness of a solution for a deterministic equation.

LEMMA 3.5.– Let a function $f \colon \mathbb{R}_+ \to \mathbb{R}$ be right-continuous everywhere on $\mathbb{R}_+$, $f(0) = 0$, and $\mathrm{var}_f(t)$ finite for all $t \in \mathbb{R}_+$, $z_0 \in \mathbb{R}$. In the class of càdlàg functions $z \colon \mathbb{R}_+ \to \mathbb{R}$, the equation

$$z(t) = z_0 + \int_0^t z(s-) \, df(s), \quad t \in \mathbb{R}_+,$$

has no more than one solution.

PROOF.– Let z be the difference of two solutions. Then

$$z(t) = \int_0^t z(s-) \, df(s), \quad t \in \mathbb{R}_+.$$

Put $g(t) := \mathrm{var}_f(t)$. Note that, by Itô's formula,

$$g(t)^n = n \int_0^t g(s-)^{n-1} dg(s) + \sum_{s \leqslant t} \left(g(s)^n - g(s-)^n - ng(s-)^{n-1}(g(s) - g(s-)) \right),$$

the summands being nonnegative. Due to this and putting $K := \sup_{s \leqslant t} |z(s)|$, we obtain, consecutively, for $r \in [0, t]$,

$$|z(r)| \leqslant \int_0^r |z(s-)| \, dg(s) \leqslant Kg(r),$$

$$|z(r)| \leqslant \int_0^r |z(s-)| \, dg(s) \leqslant K \int_0^r g(s-) \, dg(s) \leqslant Kg(r)^2/2,$$

$$|z(r)| \leqslant \int_0^r |z(s-)| \, dg(s) \leqslant \frac{K}{2} \int_0^r g(s-)^2 \, dg(s) \leqslant Kg(r)^3/6.$$

Continuing, we get $|z(r)| \leqslant Kg(r)^n/n!$. Since $g(r)^n/n! \to 0$ as $n \to \infty$, the claim follows. □

THEOREM 3.12.– Let X be a semimartingale. The equation [3.15] has a unique (up to an evanescent set) solution in the class of adapted càdlàg processes. This solution is a semimartingale, denoted by $\mathscr{E}(X)$, is given by:

$$\mathscr{E}(X) = \exp\left(X - X_0 - \frac{1}{2}\langle X^c, X^c \rangle \right) V, \qquad [3.17]$$

where V is defined in lemma 3.4.

PROOF.– Put $Y = X - X_0 - \frac{1}{2}\langle X^c, X^c \rangle$, $F(y,v) = e^y v$, $Z := F(Y,V)$. By Itô's formula, Z is a semimartingale. Since $V^c = 0$ and $Z_0 = 1$, we have

$$Z = 1 + Z_- \cdot Y + e^{Y_-} \cdot V + \frac{1}{2} Z_- \cdot \langle Y^c, Y^c \rangle + S\big(Z - Z_- - Z_- \Delta Y - e^{Y_-} \Delta V \big).$$

Since $Y^c = X^c$, the sum of the first integral ($Z_- \cdot Y$) and the third integral ($\frac{1}{2} Z_- \cdot \langle Y^c, Y^c \rangle$) is $Z_- \cdot X$. Further, since $V \in \mathscr{V}^d$, the second integral $e^{Y_-} \cdot V$ is $S(e^{Y_-} \Delta V)$, so that it can be canceled with the corresponding term in the last sum. Finally, using [3.16], we get

$$Z = e^Y V = e^{Y_- + \Delta Y} V_- (1 + \Delta X) e^{-\Delta X}$$
$$= Z_- e^{\Delta Y} (1 + \Delta X) e^{-\Delta X} = Z_- (1 + \Delta X),$$

where $Z - Z_- - Z_- \Delta Y = 0$. Resuming, we obtain $Z = 1 + Z_- \cdot X$, i.e. Z satisfies [3.15].

Assume now that an adapted càdlàg process $\widetilde{Z}$ is a solution to equation [3.15]. Then $\widetilde{Z}$ is a semimartingale. Put $\widetilde{V} = e^{-Y} \widetilde{Z} = F(-Y, \widetilde{Z})$, where Y and F are the same as before. Since $\widetilde{Z}_0 = 1$, by Itô's formula,

$$\widetilde{V} = 1 - \widetilde{V}_- \cdot Y + e^{-Y_-} \cdot \widetilde{Z} + \frac{1}{2} \widetilde{V}_- \cdot \langle Y^c, Y^c \rangle - e^{-Y_-} \cdot \langle Y^c, \widetilde{Z}^c \rangle$$
$$+ S\big(\widetilde{V} - \widetilde{V}_- + \widetilde{V}_- \Delta Y - e^{-Y_-} \Delta \widetilde{Z} \big).$$

Using propositions 3.14, 3.10 and 3.12, we obtain

$$e^{-Y_-} \cdot \widetilde{Z} = (e^{-Y_-} \widetilde{Z}_-) \cdot X = \widetilde{V}_- \cdot X,$$

$$e^{-Y_-} \cdot \langle Y^c, \widetilde{Z}^c \rangle = e^{-Y_-} \cdot \langle X^c, \widetilde{Z}_- \cdot X^c \rangle = (e^{-Y_-} \widetilde{Z}_-) \cdot \langle X^c, X^c \rangle = \widetilde{V}_- \cdot \langle X^c, X^c \rangle,$$

$$e^{-Y_-} \Delta \widetilde{Z} = e^{-Y_-} \widetilde{Z}_- \Delta X = \widetilde{V}_- \Delta X.$$

We conclude that

$$\widetilde{V} - 1 = S(\Delta\widetilde{V}),$$

i.e. $\widetilde{V} - 1 \in \mathscr{V}^d$ (more precisely, there is a version of $\widetilde{V}$ having this property). Next, $\widetilde{V} = e^{-Y}\widetilde{Z} = e^{-Y_-}e^{-\Delta X}\widetilde{Z}_-(1 + \Delta X) = \widetilde{V}_- e^{-\Delta X}(1 + \Delta X)$. Let us define the process

$$A := S\left(e^{-\Delta X}(1 + \Delta X) - 1\right).$$

Since $e^{-x}(1 + x) - 1 \sim x^2$ as $x \to 0$, it is easy to see that this process is well defined and, hence, belongs to $\mathscr{V}^d$. So we have

$$\widetilde{V} - 1 = \widetilde{V}_- \cdot A,$$

because the processes on the left and on the right belong to $\mathscr{V}^d$ and their jumps coincide. It follows from lemma 3.5 that $\widetilde{V}$ is determined uniquely (up to an evanescent set). Therefore, $\widetilde{Z} = e^Y \widetilde{V}$ is determined uniquely. $\square$

The following properties of the stochastic exponential follow from the definition, i.e. from equation [3.15].

PROPOSITION 3.15.– Let $X \in \mathscr{M}_{\text{loc}}$ (respectively $X \in \mathscr{V}$, respectively $X \in \mathscr{S}_p$, respectively X be predictable). Then, $\mathscr{E}(X) \in \mathscr{M}_{\text{loc}}$ (respectively $\mathscr{E}(X) - 1 \in \mathscr{V}$, respectively $X \in \mathscr{S}_p$, respectively $\mathscr{E}(X)$ is predictable).

PROPOSITION 3.16.– Let X be a semimartingale, ξ a $\mathscr{F}_0$-measurable random variable. The equation

$$Z = \xi + Z_- \cdot X$$

has a unique (up to an evanescent set) solution in the class of adapted càdlàg processes. This solution is the semimartingale $\xi\mathscr{E}(X)$.

EXERCISE 3.11.– Prove proposition 3.16.

HINT.– To prove the uniqueness, consider the difference of two solutions.

The following two propositions follow from lemma 3.4 and [3.17]. All relations between processes are understood up to an evanescent set.

PROPOSITION 3.17.– Let X be a semimartingale and $\Delta X > -1$. Then $\mathscr{E}(X) > 0$ and $\mathscr{E}(X)_- > 0$.

PROPOSITION 3.18.– Let X be a semimartingale, $T := \inf\{t\colon \Delta X_t = -1\}$. Then $\mathscr{E}(X) \neq 0$ on the interval $[\![0, T[\![$, and $\mathscr{E}(X)_- \neq 0$ on the interval $[\![0, T]\!]$, and $\mathscr{E}(X) = 0$ on $[\![T, \infty[\![$.

Unlike the usual exponential function, the stochastic exponential of a sum, in general, is not the product of the stochastic exponentials. However, the following statement holds.

THEOREM 3.13 (Yor's formula).– Let X and Y be semimartingales. Then

$$\mathscr{E}(X)\mathscr{E}(Y) = \mathscr{E}(X + Y + [X, Y]).$$

PROOF.– By the formula [3.12] of integration by parts and proposition 3.14, we get

$$
\begin{aligned}
\mathscr{E}(X)\mathscr{E}(Y) &= 1 + \mathscr{E}(Y)_- \cdot \mathscr{E}(X) + \mathscr{E}(X)_- \cdot \mathscr{E}(Y) + [\mathscr{E}(X), \mathscr{E}(Y)] \\
&= 1 + \big(\mathscr{E}(X)_-\mathscr{E}(Y)_-\big) \cdot X + \big(\mathscr{E}(X)_-\mathscr{E}(Y)_-\big) \cdot Y \\
&\quad + \big(\mathscr{E}(X)_-\mathscr{E}(Y)_-\big) \cdot [X, Y] \\
&= 1 + \big(\mathscr{E}(X)_-\mathscr{E}(Y)_-\big) \cdot (X + Y + [X, Y]). \qquad \square
\end{aligned}
$$

It is often useful to represent a given process Z in the form $Z_0\mathscr{E}(X)$, i.e. to express it as the stochastic exponential of another process (usually, if Z is nonnegative or strictly positive). Proposition 3.18 says that this is possible not for every semimartingale Z. We consider only the case where neither Z nor Z_- vanish.

THEOREM 3.14.– Let Z be a semimartingale such that the processes Z and Z_- do not vanish. Then there exists the process:

$$X = \frac{1}{Z_-} \cdot Z, \qquad\qquad [3.18]$$

denoted by $\mathscr{L}og\, Z$ which is called the stochastic logarithm of Z, and X is a unique semimartingale such that

$$Z = Z_0\mathscr{E}(X) \quad \text{and} \quad X_0 = 0.$$

Moreover, $\Delta X \neq -1$ and

$$\mathscr{L}og\, Z = \log\left|\frac{Z}{Z_0}\right| + \frac{1}{2Z_-^2} \cdot \langle Z^c, Z^c\rangle - S\left(\log\left|1 + \frac{\Delta Z}{Z_-}\right| - \frac{\Delta Z}{Z_-}\right). \qquad [3.19]$$

PROOF.– Put $T_n := \inf\{t\colon |Z_t| < 1/n\}$. Since Z and Z_- do not vanish, we have $T_n \uparrow \infty$. Since $|Z_-| \geqslant 1/n$ on $]\!]0, T_n]\!]$, the process $1/Z_-$ is locally bounded and the process X in [3.18] is well defined. Moreover, $\Delta X = \Delta Z / Z_- \neq -1$. Using proposition 3.14, we get

$$Z_- \cdot X = Z - Z_0,$$

i.e. $Z = Z_0 \mathcal{E}(X)$ due to proposition 3.16. If X is another semimartingale satisfying this relation and $X_0 = 0$, then $Z = Z_0 + Z_- \cdot X$, and it follows from proposition 3.14 that

$$\frac{1}{Z_-} \cdot Z = \left(Z \cdot \frac{1}{Z_-}\right) \cdot X = X.$$

In order to prove [3.19], we can use Itô's formula with a twice continuously differentiable function F_n such that $F_n(x) = \log|x|$ on the set $\{x \in \mathscr{R}\colon |x| \geqslant 1/n\}$. Using the arguments preceding [3.10], we can prove [3.19] on the stochastic interval $[\![0, T_n]\!]$ for every n. □

EXERCISE 3.12.– Give a full proof of [3.19].

3.4. Stochastic integrals with respect to semimartingales: the general case

In this section, we will study stochastic integrals with respect to semimartingales in the general case. That is, integrands are predictable but not locally bounded as in section 3.2. In order to avoid any misunderstandings and for clarity, we will denote the pathwise Lebesgue–Stieltjes integral defined in theorem 2.11 by $H \overset{s}{\cdot} X$ and the stochastic integral with respect to local martingales, see definition 3.3, by $H \overset{m}{\cdot} X$; the symbol $H \cdot X$ is reserved for stochastic integrals with respect to semimartingales.

The idea how to define the integral is the same: to take a decomposition [3.7] and to define $H \cdot X$ as $H \overset{m}{\cdot} M + H \overset{s}{\cdot} A$, Of course, we should assume that $H \in L^1_{\mathrm{loc}}(M) \cap L_{\mathrm{var}}(A)$. The main difficulty of this approach is that the class $L^1_{\mathrm{loc}}(M) \cap L_{\mathrm{var}}(A)$ depends on the choice of M and A in [3.7]. Indeed, if X is a semimartingale from $\mathscr{M}_{\mathrm{loc}} \cap \mathscr{V}$, we may consider [3.7] with $M = X$ and $A = 0$ (the canonical decomposition) or, conversely, $M = 0$ and $A = X$. In the first case, $L^1_{\mathrm{loc}}(M) \cap L_{\mathrm{var}}(A) = L^1_{\mathrm{loc}}(X)$, and, in the second case, $L^1_{\mathrm{loc}}(M) \cap L_{\mathrm{var}}(A) = L_{\mathrm{var}}(X)$. However, we know from examples 2.3 and 3.2 that any of these classes does not contain another one, in general. The first of these examples also shows that if $X \in \mathscr{S}_p$, then the class $L^1_{\mathrm{loc}}(M) \cap L_{\mathrm{var}}(A)$ with arbitrary M and A satisfying [3.7] may contain elements which are not in the class with M and A from the canonical decomposition.

DEFINITION 3.9.– Let X be a semimartingale. A predictable process H is called X-*integrable*, if there is a decomposition $X = X_0 + M + A$, $M \in \mathcal{M}_{\text{loc}}$, $A \in \mathcal{V}$, such that $H \in L^1_{\text{loc}}(M) \cap L_{\text{var}}(A)$. In this case, the *stochastic integral* $H \cdot X$ is defined by

$$H \cdot X := H \overset{m}{\cdot} M + H \overset{s}{\cdot} A.$$

The class of all X-integrable processes is denoted by $L(X)$.

It is clear that, if $X \in \mathcal{S}$ and $H \in L(X)$, then $H \cdot X$ is a semimartingale and $H \cdot X_0 = 0$.

Obviously, any locally bounded predictable process X belongs to $L(X)$ for every X, and definition 3.8 is a special case of definition 3.9.

PROPOSITION 3.19.– Definition 3.9 of the stochastic integral is correct.

PROOF.– Let $X = X_0 + M + A = X_0 + M' + A'$ with $M, M' \in \mathcal{M}_{\text{loc}}$, $A, A' \in \mathcal{V}$, $H \in L^1_{\text{loc}}(M) \cap L^1_{\text{loc}}(M') \cap L_{\text{var}}(A) \cap L_{\text{var}}(A')$. Then $M - M' = A' - A \in \mathcal{M}_{\text{loc}} \cap \mathcal{V}$, $H \in L^1_{\text{loc}}(M - M') \cap L_{\text{var}}(A' - A)$, and by theorem 3.5 (1),

$$H \overset{m}{\cdot} (M - M') = H \overset{s}{\cdot} (A' - A).$$

Therefore,

$$H \overset{m}{\cdot} M + H \overset{s}{\cdot} A = H \overset{m}{\cdot} M' + H \overset{s}{\cdot} A'. \qquad \square$$

The next proposition follows trivially from definition 3.9 and proposition 3.19. It shows that definition 3.9 of the stochastic integral with respect to semimartingales includes the definition of the pathwise Lebesgue–Stieltjes integral from theorem 2.11 and the definition of the stochastic integral with respect to local martingales, see definition 3.3, as special cases.

PROPOSITION 3.20.–

1) Let $X \in \mathcal{V}$. Then $L_{\text{var}}(X) \subseteq L(X)$ and

$$H \overset{s}{\cdot} X = H \cdot X \quad \text{for every } H \in L_{\text{var}}(X).$$

2) Let $X \in \mathcal{M}_{\text{loc}}$. Then $L^1_{\text{loc}}(X) \subseteq L(X)$ and

$$H \overset{m}{\cdot} X = H \cdot X \quad \text{for every } H \in L^1_{\text{loc}}(X).$$

In view of this proposition, there is no need to use symbols $\overset{s}{\cdot}$ and $\overset{m}{\cdot}$.

Now we state properties of the integral introduced in definition 3.9 which follow easily from the definition and from corresponding properties of the stochastic integral with respect to local martingales and the pathwise Lebesgue–Stieltjes integral.

PROPOSITION 3.21.– Let $X, Y \in \mathscr{S}$, $H \in L(X) \cap L(Y)$, $\alpha, \beta \in \mathbb{R}$. Then, $H \in L(\alpha X + \beta Y)$ and

$$H \cdot (\alpha X + \beta Y) = \alpha(H \cdot X) + \beta(H \cdot Y).$$

PROPOSITION 3.22.– Let X be a semimartingale and $H \in L(X)$. Then, $H \in L^1_{\mathrm{loc}}(X^c)$ and $(H \cdot X)^c = H \cdot X^c$.

PROPOSITION 3.23.– Let X be a semimartingale, $H \in L(X)$, and let T be a stopping time. Then, $H \in L(X^T)$, $H\mathbb{1}_{\rrbracket 0,T\rrbracket} \in L(X)$ and

$$(H \cdot X)^T = H \cdot X^T = (H\mathbb{1}_{\rrbracket 0,T\rrbracket}) \cdot X.$$

PROPOSITION 3.24.– Let X be a semimartingale and $H \in L(X)$. Then

$$\Delta(H \cdot X) = H\Delta X.$$

PROPOSITION 3.25.– Let X be a semimartingale and $H \in L(X)$. Then, for every semimartingale Y, we have $H \in L_{\mathrm{var}}([X,Y])$ and

$$[H \cdot X, Y] = H \cdot [X, Y].$$

EXERCISE 3.13.– Prove propositions 3.21–3.25.

As regards the linearity of $H \cdot X$ in H and the "associative" property $K \cdot (H \cdot X) = (KH) \cdot X$, some preparation is needed.

We have already observed that, for a special semimartingale X with the canonical decomposition $X = X_0 + M + A$, the class $L^1_{\mathrm{loc}}(M) \cap L_{\mathrm{var}}(A)$ may be more narrow than $L(X)$ as in example 2.3. The following important theorem characterizes this situation and generalizes statement (3) in theorem 3.5.

THEOREM 3.15.– Let X be a special semimartingale with the canonical decomposition $X = X_0 + M + A$ and $H \in L(X)$. A necessary and sufficient condition for $H \cdot X$ to be a special semimartingale is that $H \in L^1_{\mathrm{loc}}(M) \cap L_{\mathrm{var}}(A)$.

Note that if X is a predictable semimartingale and $H \in L(X)$, then $H \cdot X$ is also predictable by proposition 3.24. By corollary 3.8, X and $H \cdot X$ are special semimartingales. Therefore, it holds

COROLLARY 3.10.–

1) Let X be a predictable semimartingale with the canonical decomposition $X = X_0 + M + A$. Then, $L(X) = L^1_{\text{loc}}(M) \cap L_{\text{var}}(A)$.

2) Let M be a continuous local martingale. Then, $L(M) = L^1_{\text{loc}}(M)$.

3) Let A be a predictable process with finite variation. Then, $L(A) = L_{\text{var}}(A)$.

Before we turn to the proof of theorem 3.15, we state and prove an intermediate version of the "associative" property with a locally bounded K.

LEMMA 3.6.– Let X be a semimartingale, $H \in L(X)$, K a locally bounded predictable process. Then $KH \in L(X)$ and $K \cdot (H \cdot X) = (KH) \cdot X$.

PROOF.– Let a decomposition $X = X_0 + M + A$, $M \in \mathcal{M}_{\text{loc}}$, $A \in \mathcal{V}$, be such that $H \in L^1_{\text{loc}}(M) \cap L_{\text{var}}(A)$. Note that $H \cdot M \in \mathcal{M}_{\text{loc}}$ and $K \in L^1_{\text{loc}}(H \cdot M)$ because K is locally bounded. Hence, by proposition 3.7,

$$KH \in L^1_{\text{loc}}(M) \quad \text{and} \quad K \cdot (H \cdot M) = (KH) \cdot M.$$

Similarly, by theorem 2.12,

$$KH \in L_{\text{var}}(A) \quad \text{and} \quad K \cdot (H \cdot A) = (KH) \cdot A.$$

Therefore, $KH \in L(X)$ and

$$
\begin{aligned}
(KH) \cdot X &= (KH) \cdot M + (KH) \cdot A \\
&= K \cdot (H \cdot M) + K \cdot (H \cdot A) = K \cdot (H \cdot X). \qquad \square
\end{aligned}
$$

PROOF OF THEOREM 3.15.– The sufficiency is obvious. Indeed, $H \cdot M \in \mathcal{M}_{\text{loc}}$ for $H \in L^1_{\text{loc}}(M)$ by definition 3.3 and the process $H \cdot A$ belongs to $\mathcal{V}$ and is predictable for $H \in L_{\text{var}}(A)$ by theorem 2.12.

Let us prove the necessity. Let $H \cdot X$ be a special semimartingale. Write its canonical decomposition as

$$H \cdot X = N + B,$$

where $N \in \mathcal{M}_{\text{loc}}$, $B \in \mathcal{V}$, and B is predictable. Put

$$K := \frac{1}{1 + |H|}, \quad J := \frac{1}{K}.$$

It is obvious that K and J are strictly positive predictable processes, K and KH are bounded.

By proposition 3.8, $(KH) \cdot X$ and $K \cdot (H \cdot X)$ are special semimartingales with the canonical decompositions

$$(KH) \cdot X = (KH) \cdot M + (KH) \cdot A,$$

$$K \cdot (H \cdot X) = K \cdot N + K \cdot B.$$

However, by lemma 3.6, $(KH) \cdot X = K \cdot (H \cdot X)$. Since the canonical decomposition is unique, we have:

$$(KH) \cdot M = K \cdot N, \quad (KH) \cdot A = K \cdot B. \qquad [3.20]$$

Since $JK = 1$, we have $JK \in L^1_{\text{loc}}(N)$. By proposition 3.7, we get $J \in L^1_{\text{loc}}(K \cdot N)$. Taking into account the first equality in [3.20], we have $J \in L^1_{\text{loc}}((KH) \cdot M)$. Applying proposition 3.7 once again, we obtain $H = JKH \in L^1_{\text{loc}}(M)$.

Similarly, $J \in L_{\text{var}}(K \cdot B)$, hence the second equality in [3.20] implies $J \in L_{\text{var}}((KH) \cdot A)$, from which $H = JKH \in L_{\text{var}}(A)$ follows. $\qquad \square$

COROLLARY 3.11.– Let $M \in \mathscr{M}_{\text{loc}}$ and $H \in L(M)$. A necessary and sufficient condition for $H \cdot M$ to be a local martingale is that $H \in L^1_{\text{loc}}(M)$.

PROOF.– The sufficiency follows from definition 3.3, and the necessity comes from theorem 3.15. $\qquad \square$

In the case where X is a predictable process with finite variation, the assertion of theorem 3.15 reduces to corollary 3.10 (3). But we can prove the following interesting result for processes with finite variation that are not necessarily predictable, which generalizes statement (2) in theorem 3.5.

THEOREM 3.16.– Let $A \in \mathscr{V}$ and $H \in L(A)$. It is necessary and sufficient for $H \cdot A \in \mathscr{V}$ that $H \in L_{\text{var}}(A)$.

PROOF.– Again, it is enough to check the necessity. Let $A \in \mathscr{V}$, $H \in L(A)$ and $H \cdot A \in \mathscr{V}$. It follows from the assumption $H \in L(A)$ that there is a decomposition $A = M + B$ such that $M \in \mathscr{M}_{\text{loc}}$, $B \in \mathscr{V}$ and $H \in L^1_{\text{loc}}(M) \cap L_{\text{var}}(B)$. Then,

$$H \cdot A = H \cdot M + H \cdot B.$$

Hence, $M = A - B \in \mathscr{V}$ and $H \cdot M = H \cdot A - H \cdot B \in \mathscr{V}$. By theorem 3.5 (2), $H \in L_{\text{var}}(M)$, and we obtain $H \in L_{\text{var}}(M + B) = L_{\text{var}}(A)$. $\qquad \square$

Let us call a set $D \subset \Omega \times \mathbb{R}_+$ *discrete* if, for almost all ω, the set $\{s \in \mathbb{R}_+ : (\omega, s) \in D, s \leqslant t\}$ is finite for all $t \in \mathbb{R}_+$. If X is a càdlàg process, then, for every $a > 0$, the set $\{|\Delta X| > a\}$ is discrete.

Let X be an adapted càdlàg process and D an optional discrete set. Then the process $S(\Delta X \mathbb{1}_D) \in \mathcal{V}$ is well defined. Put $X^D := X - S(\Delta X \mathbb{1}_D)$. Since $\Delta X^D = \mathbb{1}_{(\Omega \times \mathbb{R}_+) \setminus D} \Delta X$, by theorem 3.7, if X is a semimartingale, then X^D is a special semimartingale provided

$$D \supseteq \{|\Delta X| > 1\}.$$

LEMMA 3.7.– Let X be a semimartingale and $H \in L(X)$.

1) The set

$$D_0 := \{|\Delta X| > 1\} \cup \{|H\Delta X| > 1\}$$

is an optional discrete set.

2) Let D be an optional discrete set containing D_0, $X^D = X_0 + M + B$ the canonical decomposition of the special semimartingale X^D. Put $A := B + S(\Delta X \mathbb{1}_D)$. Then, $H \in L^1_{\text{loc}}(M) \cap L_{\text{var}}(A)$. Moreover, $(H \cdot X)^D$ is a special semimartingale, and $H \cdot M$ is the local martingale in its canonical decomposition.

Thus, the lemma allows us to construct explicitly a decomposition $X = X_0 + M + A$, $M \in \mathcal{M}_{\text{loc}}$, $A \in \mathcal{V}$, such that $H \in L^1_{\text{loc}}(M) \cap L_{\text{var}}(A)$ (assuming that $H \in L(X)$).

PROOF.– Put $Y := H \cdot X$.

1) By proposition 3.24, $\{|H\Delta X| > 1\} = \{|\Delta Y| > 1\}$ up to an evanescent set. The claim follows.

2) Since the set D is discrete and $\Delta Y = H\Delta X$, it is evident that $H \in L_{\text{var}}(S(\Delta X \mathbb{1}_D))$ and $H \cdot S(\Delta X \mathbb{1}_D) = S(\Delta Y \mathbb{1}_D)$. Therefore, by proposition 3.21, $H \in L(X^D)$ and $H \cdot X^D = Y^D$. Since $D \supseteq D_0$, X^D and Y^D are special semimartingales. Theorem 3.15 allows us to conclude that $H \in L^1_{\text{loc}}(M) \cap L_{\text{var}}(B)$. Moreover, $Y^D = H \cdot M + H \cdot B$ is the canonical decomposition of Y^D. That $H \in L_{\text{var}}(A)$ follows from the definition of A. $\square$

Now we are in a position to prove the remaining properties of the stochastic integral.

PROPOSITION 3.26.– Let X be a semimartingale, $H, K \in L(X)$, $\alpha, \beta \in \mathbb{R}$. Then $\alpha H + \beta K \in L(X)$ and

$$(\alpha H + \beta K) \cdot X = \alpha(H \cdot X) + \beta(K \cdot X).$$

PROOF.– Applying lemma 3.7 with

$$D := \{|\Delta X| > 1\} \cup \{|H\Delta X| > 1\} \cup \{|K\Delta X| > 1\},$$

we get a decomposition $X = X_0 + M + A$, $M \in \mathcal{M}_{\text{loc}}$, $A \in \mathcal{V}$, such that $H, K \in L^1_{\text{loc}}(M) \cap L_{\text{var}}(A)$. Now, using the linearity of stochastic integrals with respect to local martingales and Lebesgue–Stieltjes integrals, we obtain

$$\alpha H + \beta K \in L^1_{\text{loc}}(M) \cap L_{\text{var}}(A),$$

$$(\alpha H + \beta K) \cdot M = \alpha(H \cdot M) + \beta(K \cdot M),$$

$$(\alpha H + \beta K) \cdot A = \alpha(H \cdot A) + \beta(K \cdot A).$$

The claim follows. □

PROPOSITION 3.27.– Let X be a semimartingale, $H \in L(X)$, and let K be a predictable process. Then

$$K \in L(H \cdot X) \Leftrightarrow KH \in L(X),$$

and, in this case,

$$K \cdot (H \cdot X) = (KH) \cdot X.$$

PROOF.– Assume that $K \in L(H \cdot X)$. Denote $Y := H \cdot X$, $Z := K \cdot Y$, and put

$$D := \{|\Delta X| > 1\} \cup \{|H\Delta X| > 1\} \cup \{|KH\Delta X| > 1\}.$$

Since $\{|H\Delta X| > 1\} = \{|\Delta Y| > 1\}$ and $\{|KH\Delta X| > 1\} = \{|K\Delta Y| > 1\} = \{|\Delta Z| > 1\}$ (up to an evanescent set), D is an optional discrete set. Therefore, we can apply lemma 3.7 to the pairs (X, H) and (Y, K) with this set D and to get decompositions $X = X_0 + M + A$ and $Y = N + B$, $M, N \in \mathcal{M}_{\text{loc}}$, $A, B \in \mathcal{V}$, for which $H \in L^1_{\text{loc}}(M) \cap L_{\text{var}}(A)$, $K \in L^1_{\text{loc}}(N) \cap L_{\text{var}}(B)$. Moreover, by lemma 3.7, M is the local martingale in the canonical decomposition of the special semimartingale X^D, N is the local martingale in the canonical decomposition of the special semimartingale Y^D, and $N = H \cdot M$. Hence,

$B = H \cdot A$. It follows from the associative properties of stochastic integrals with respect to local martingales and Lebesgue–Stieltjes integrals that

$$KH \in L^1_{\mathrm{loc}}(M) \cap L_{\mathrm{var}}(A),$$

$$K \cdot N = K \cdot (H \cdot M) = (KH) \cdot M,$$

$$K \cdot B = K \cdot (H \cdot A) = (KH) \cdot A.$$

This implies $KH \in L(X)$ and $K \cdot (H \cdot X) = (KH) \cdot X$.

Assume that $KH \in L(X)$. Under this assumption, the set D introduced above is also discrete. Thus, by lemma 3.7, if $X = X_0 + M + A$ is the same decomposition, we have $H \in L^1_{\mathrm{loc}}(M) \cap L_{\mathrm{var}}(A)$ and $KH \in L^1_{\mathrm{loc}}(M) \cap L_{\mathrm{var}}(A)$. Hence, $K \in L^1_{\mathrm{loc}}(H \cdot M) \cap L_{\mathrm{var}}(H \cdot A)$, where $K \in L(H \cdot X)$. □

3.5. σ-martingales

DEFINITION 3.10.– A semimartingale X is called a σ-martingale if there is a sequence of predictable sets D_n such that $D_1 \subseteq \cdots \subseteq D_n \subseteq \ldots$, $\bigcup_n D_n = \Omega \times \mathbb{R}_+$ (up to an evanescent set), and the processes $\mathbb{1}_{D_n} \cdot X$ are uniformly integrable martingales for all n. The class of all σ-martingales will be denoted by $\mathscr{M}_\sigma$.

It is clear that any local martingale is a σ-martingale: put $D_n = \mathbb{1}_{[0,T_n]}$, where $\{T_n\}$ is a localizing sequence from the definition of a local martingale. Moreover, as follows from the definition, X is a σ-martingale if and only if $X - X_0$ is a σ-martingale. On the other hand, our definition of a local martingale implies that X_0 is integrable. So, it is easy to construct examples of σ-martingales X that are not local martingales: just take a local martingale M and put $X = \xi + M$, where ξ is non-integrable $\mathscr{F}_0$-measurable random variable. But there are also σ-martingales X with $X_0 = 0$, which are not local martingales, see examples below.

It follows from the definition that the class of σ-martingales is stable under stopping and scalar multiplication. That this class is stable under addition is less obvious.

THEOREM 3.17.– Let X be a semimartingale. The following statements are equivalent:

1) there is a sequence of predictable sets D_n such that $D_1 \subseteq \cdots \subseteq D_n \subseteq \ldots$, $\bigcup_n D_n = \Omega \times \mathbb{R}_+$ (up to an evanescent set) and $\mathbb{1}_{D_n} \cdot X \in \mathscr{M}_{\mathrm{loc}}$ for all n;

2) X is a σ-martingale;

3) there is a sequence of predictable sets D_n such that $D_1 \subseteq \cdots \subseteq D_n \subseteq \dots$, $\bigcup_n D_n = \Omega \times \mathbb{R}_+$ and $\mathbb{1}_{D_n} \cdot X \in \mathscr{H}^1$ for all n;

4) there is a process $G \in L(X)$ such that $G \neq 0$ identically and $G \cdot X \in \mathscr{M}_{\mathrm{loc}}$;

5) there is a process $J \in L(X)$ such that $J > 0$ identically and $J \cdot X \in \mathscr{H}^1$;

6) there are processes $M \in \mathscr{M}_{\mathrm{loc}}$ and $H \in L(M)$ such that $X = X_0 + H \cdot M$;

7) there are processes $N \in \mathscr{H}^1$ and $K \in L(N)$ such that $K > 0$ identically and $X = X_0 + K \cdot N$.

PROOF.– Note that implications (3)$\Rightarrow$(2)$\Rightarrow$(1), (5)$\Rightarrow$(4), (7)$\Rightarrow$(6) are obvious. If (4) holds, then put $M := G \cdot X$ and $H := 1/G$. Then $H \in L(M)$ and $H \cdot M = X - X_0$ by proposition 3.27, i.e. (6) holds. Similar arguments show that (5) and (7) are equivalent. Therefore, it is enough to prove implications (1)$\Rightarrow$(3), (3)$\Rightarrow$(5) and (6)$\Rightarrow$(1).

We start with the last one. Let $M \in \mathscr{M}_{\mathrm{loc}}$, $H \in L(M)$ and $X = X_0 + H \cdot M$. Put $D_n := \{|H| \leqslant n\}$, then D_n are predictable and increasing to $\Omega \times \mathbb{R}_+$. Moreover, by proposition 3.27, for every n,

$$\mathbb{1}_{D_n} \cdot X = H \mathbb{1}_{\{|H| \leqslant n\}} \cdot M,$$

and the process on the right is a local martingale because the integrand is bounded.

Now assume (1) and let us prove (3). Put $M^n := \mathbb{1}_{D_n} \cdot X$. Since $M^n \in \mathscr{M}_{\mathrm{loc}}$, by theorem 2.7, for every n, there is a localizing sequence $\{T_{n,p}\}_{p \in \mathbb{N}}$ of stopping times such that $(M^n)^{T_{n,p}} \in \mathscr{H}^1$ for all p. By lemma 2.1 (2) (with $T_n = \infty$), there is a localizing sequence $\{S_n\}$ of stopping times such that, for every n, $S_n \leqslant T_{n,p_n}$ for some p_n. Hence, $(M^n)^{S_n} \in \mathscr{H}^1$. Since $S_n \uparrow \infty$ a.s., the set $(\Omega \times \mathbb{R}_+) \setminus \bigcup_n (D_n \cap [\![0, S_n]\!])$ is evanescent (and predictable); we denote it by N. Now put $C_n := (D_n \cap [\![0, S_n]\!]) \cup N$. It is clear that the sets C_n are predictable and increasing to $\Omega \times \mathbb{R}_+$. Moreover, since N is evanescent, by proposition 3.27, for every n,

$$\mathbb{1}_{C_n} \cdot X = \mathbb{1}_{D_n \cap [\![0, S_n]\!]} \cdot X = \mathbb{1}_{[\![0, S_n]\!]} \cdot (\mathbb{1}_{D_n} \cdot X) = \mathbb{1}_{[\![0, S_n]\!]} \cdot M^n \in \mathscr{H}^1.$$

Finally, we prove implication (3)$\Rightarrow$(5). Put $M^n := \mathbb{1}_{D_n} \cdot X$; we have $M^n \in \mathscr{H}^1$ for every n. Choose a sequence of numbers $\{\alpha_n\}$ such that $0 < \alpha_n \leqslant 1/2$, $\alpha_n \|M^n\|_{\mathscr{H}^1} \leqslant 2^{-n}$, and $\alpha_{n+1} \leqslant \alpha_n/2$ for all n. Put

$$J := \sum_{n=1}^{\infty} \alpha_n \mathbb{1}_{D_n}.$$

It is clear that $0 < J \leqslant 1$. In particular, $J \in L(X)$ and the integral $Y := J \cdot X \in \mathscr{S}$ is defined. It remains to prove that Y is a martingale from $\mathscr{H}^1$.

By proposition 3.27,

$$\mathbb{1}_{D_k} \cdot Y = (J\mathbb{1}_{D_k}) \cdot X = \sum_{n=1}^{k} \alpha_n (\mathbb{1}_{D_n} \cdot X) + \left(\sum_{n=k+1}^{\infty} \alpha_n \right) \mathbb{1}_{D_k} \cdot X$$

$$= \sum_{n=1}^{k} \alpha_n M^n + \left(\sum_{n=k+1}^{\infty} \alpha_n \right) M^k.$$

Since $0 < \sum_{n=k+1}^{\infty} \alpha_n \leqslant \alpha_k$, the second term on the right converges to 0 in $\mathcal{H}^1$ as $k \to \infty$. On the other hand, since the series $\sum \alpha_n \|M^n\|_{\mathcal{H}^1}$ converges and the space $\mathcal{H}^1$ is complete (theorem 2.6), the series $\sum_{n=1}^{\infty} \alpha_n M^n$ converges in $\mathcal{H}^1$. Denote the sum of this series by N. Then, $\mathbb{1}_{D_k} \cdot Y \to N$ in $\mathcal{H}^1$ as $k \to \infty$. In particular,

$$\sup_{s \in \mathbb{R}_+} |\mathbb{1}_{D_k} \cdot Y_s - N_s| \xrightarrow{P} 0, \quad k \to \infty.$$

However, by theorem 3.9,

$$\sup_{s \leqslant t} |\mathbb{1}_{D_k} \cdot Y_s - Y_s| \xrightarrow{P} 0, \quad k \to \infty.$$

for every $t \in \mathbb{R}_+$. Hence, trajectories of N and Y coincide on $[0, t]$ with probability one, for every t. Therefore, N and Y are indistinguishable, and $Y \in \mathcal{H}^1$. $\qquad \square$

EXERCISE 3.14.– Prove that the sum of two σ-martingales is a σ-martingale.

The equivalence of statement (6) of the theorem to the definition of a σ-martingale allows us to provide an example of a σ-martingale starting from 0, which is not a local martingale, namely, the process X in example 2.3. Here is another example.

EXAMPLE 3.3.– Let there be given a complete probability space $(\Omega, \mathscr{F}, \mathsf{P})$ with independent random variables $\xi_1, \ldots, \xi_n, \ldots$ on it. We assume that

$$\mathsf{P}(\xi_n = \pm n) = 1/(2n^2), \quad \mathsf{P}(\xi_n = 0) = 1 - 1/n^2.$$

Put $t_n = n/(n+1)$, $\mathscr{F}_t := \sigma\{\xi_1, \ldots, \xi_n, \ldots : t_n \leqslant t\} \vee \sigma\{\mathscr{N}\}$, where $\mathscr{N}$ consists of P-null sets from $\mathscr{F}$. By the Borel–Cantelli lemma, for almost all ω, $\xi_n(\omega) \neq 0$ only for a finite number of n, and the series $\sum_{n=1}^{\infty} \xi_n$ converges. For such ω, put

$$X_t := \sum_{n : t_n \leqslant t} \xi_n, \quad t \in \mathbb{R}_+,$$

while, for other ω, put $X_t = 0$ for all $t \in \mathbb{R}_+$. Obviously, $X \in \mathcal{V}$, because all trajectories are piecewise constant (with a finite number of pieces). Next, put $D_n := \Omega \times ([0, t_n] \cup [1, \infty[)$. Then $\mathbb{1}_{D_n} \cdot X$ is, obviously, a bounded martingale. Hence, X is a σ-martingale.

Assume that X is a local martingale. Then, by lemma 2.6, $X \in \mathcal{A}_{\mathrm{loc}}$, i.e. $A := \mathrm{Var}\,(X)$ is a locally integrable increasing process. Denote the compensator of A by $\widetilde{A}$. Then, for every n, $\widetilde{A}^{t_n}$ is the compensator of the process A^{t_n}. Since for almost all ω,

$$A_t^{t_n} = \sum_{k:\, k \leqslant n,\, t_k \leqslant t} |\xi_k|, \quad t \in \mathbb{R}_+,$$

A^{t_n} is a process with independent increments on our stochastic basis. It follows from exercise 2.6 that

$$\widetilde{A}_t^{t_n} = \sum_{k:\, k \leqslant n,\, t_k \leqslant t} E|\xi_k|, \quad t \in \mathbb{R}_+.$$

Hence, for every n,

$$\widetilde{A}_1 \geqslant \widetilde{A}_{t_n} = \widetilde{A}_{t_n}^{t_n} = \sum_{k=1}^{n} \frac{1}{k}.$$

Thus, $\widetilde{A}_1 = \infty$ a.s. This contradiction shows that X is not a local martingale.

This example is especially interesting because X is a process with independent increments (on the stochastic basis under consideration). In this connection, let us mention the following fact (without proof): if a local martingale is a process with independent increments, then it is a martingale (all the notions are with respect to the same stochastic basis), see [SHI 02].

We know that the integral (in the sense of definition 3.9) with respect to a local martingale need not be a local martingale, and the integral with respect to a process with finite variation is not necessarily a process with finite variation. It turns out that the integral with respect to a σ-martingale is always a σ-martingale.

THEOREM 3.18.– Let $X \in \mathcal{M}_\sigma$ and $H \in L(X)$. Then $H \cdot X \in \mathcal{M}_\sigma$.

PROOF.– By theorem 3.17, $X = X_0 + K \cdot M$, where $M \in \mathcal{M}_{\mathrm{loc}}$ and $K \in L(M)$. By theorem 3.27, $HK \in L(M)$ and $H \cdot X = (HK) \cdot M$. Applying theorem 3.17 again, we get $H \cdot X \in \mathcal{M}_\sigma$. □

THEOREM 3.19.– Let X be a σ-martingale. The following statements are equivalent:

1) $X - X_0$ is a local martingale;

2) X is a special semimartingale.

PROOF.– Implication (1)$\Rightarrow$(2) is obvious (and does not use the assumption that X is a σ-martingale). Let X be both a σ-martingale and a special semimartingale. By theorem 3.17, $X = X_0 + H \cdot M$, where $M \in \mathcal{M}_{\mathrm{loc}}$ and $H \in L(M)$. Hence, $H \cdot M$ is a special semimartingale, and, by theorem 3.15, $H \in L^1_{\mathrm{loc}}(M)$ and $H \cdot M \in \mathcal{M}_{\mathrm{loc}}$. $\square$

In view of the above theorem, all conditions that are sufficient for a semimartingale to be a special semimartingale (see theorems 3.6 and 3.7), are sufficient for a σ-martingale starting from zero to be a local martingale. We mention only a few cases.

COROLLARY 3.12.– Let X be a σ-martingale. Assume that X is locally bounded or has bounded jumps. Then $X - X_0$ is a local martingale.

COROLLARY 3.13.– (1) $X \in \overline{\mathcal{M}}$ if and only if $X \in \mathcal{M}_\sigma \cap (DL)$.

(2) $X \in \mathcal{M}$ if and only if $X \in \mathcal{M}_\sigma \cap (D)$.

PROOF.– The assertions "only if" are evident due to corollary 2.4. Let $X \in \mathcal{M}_\sigma \cap (DL)$. Note that the random variable X_0 is integrable. Put $T_n := \inf\{t: |X_t| > n\} \wedge n$. Then $X^*_{T_n} \leqslant n + |X_{T_n}|$, while a random variable X_{T_n} is integrable because $X \in (DL)$. Hence, $(X - X_0)^* \in \mathscr{A}^+_{\mathrm{loc}}$, hence $X \in \mathscr{S}_p$ by theorem 3.6 and $X \in \mathcal{M}_{\mathrm{loc}}$ by theorem 3.19. We complete the proof by applying theorem 2.8. $\square$

Let us remark that all sufficient conditions mentioned before corollary 3.12 are "two-sided", i.e. constrain equally, say, big positive jumps and big (in absolute value) negative jumps. This is not surprising if we want to check that a semimartingale is a special semimartingale. However, if we deal with a σ-martingale, we can find "one-sided" necessary and sufficient conditions.

THEOREM 3.20 (Ansel–Stricker).– Let X be a σ-martingale, $X_0 = 0$. Then, X is a local martingale if and only if:

$$\left\{(\Delta X)^-\right\}^* \in \mathscr{A}^+_{\mathrm{loc}}. \tag{3.21}$$

PROOF.– Since $0 \leqslant (\Delta X)^- \leqslant |\Delta X|$, the necessity follows from lemma 2.7.

Let us prove the sufficiency. Let X be a σ-martingale, $X_0 = 0$, and let [3.21] hold. We start with a special case, where we can see a central idea of the proof. The general case will be reduced to the special case.

Namely, let X be represented as

$$X = H \cdot M, \quad M \in \mathcal{M}_{\text{loc}} \cap \mathcal{V}, \quad H \in L_{\text{var}}(M), \quad H > 0.$$

Let $M = B - C$ be the decomposition of $M \in \mathcal{V}$ from proposition 2.3. Since $M \in \mathcal{A}_{\text{loc}}$ (lemma 2.6), we have $B, C \in \mathcal{A}_{\text{loc}}^+$. Their compensators are denoted by $\widetilde{B}$ and $\widetilde{C}$, respectively. Since $B - C \in \mathcal{M}_{\text{loc}}$, we have $\widetilde{B} = \widetilde{C}$.

Next, $L_{\text{var}}(M) = L_{\text{var}}(B) \cap L_{\text{var}}(C)$ and

$$(\Delta X)^- = (H\Delta M)^- = H(\Delta M)^- = H\Delta C = \Delta(H \cdot C).$$

Since $H \cdot C \in \mathcal{V}^+$ and the process $(H \cdot C)_-$ is locally bounded, it follows from [3.21] and the last relation that $H \cdot C \in \mathcal{A}_{\text{loc}}^+$. theorem 2.21 (2) allows us to conclude that $H \cdot \widetilde{C} \in \mathcal{A}_{\text{loc}}^+$. Hence, $H \cdot \widetilde{B} \in \mathcal{A}_{\text{loc}}^+$. It follows from theorem 2.21 (3) that $H \cdot B \in \mathcal{A}_{\text{loc}}^+$. Thus, $H \cdot M \in \mathcal{A}_{\text{loc}}$, hence, $X = H \cdot M \in \mathcal{M}_{\text{loc}}$ (proposition 2.10 (4)).

We now turn to the general case. By theorem 3.17, there are processes $M \in \mathcal{M}_{\text{loc}}$ and $H \in L(M)$ such that $H > 0$ and $X = H \cdot M$. By the definition of the stochastic integral with respect to semimartingales, there is a decomposition $M = M_0 + N + A$, $N \in \mathcal{M}_{\text{loc}}, A \in \mathcal{V}$, such that $H \in L^1_{\text{loc}}(N) \cap L_{\text{var}}(A)$ and then $X = H \cdot N + H \cdot A$. Then, obviously, $A \in \mathcal{M}_{\text{loc}} \cap \mathcal{V}$ and $Y := H \cdot A = X - H \cdot N \in \mathcal{M}_\sigma$. Finally, for every Y, we have [3.21]:

$$\left\{(\Delta Y)^-\right\}^* = \left\{(\Delta X - H\Delta N)^-\right\}^* \leqslant \left\{(\Delta X)^-\right\}^* + \left\{(H\Delta N)^+\right\}^* \in \mathcal{A}_{\text{loc}}^+,$$

where we have used the inequalities $(x - y)^- \leqslant x^- + y^+$ for real x, y, relation [3.21] (for X) and also relation $(H\Delta N)^* \in \mathcal{A}_{\text{loc}}^+$, which follows from lemma 2.7 due to the fact that $H\Delta N = \Delta(H \cdot N)$, and $H \cdot N \in \mathcal{M}_{\text{loc}}$. According to the special case considered above, we have $Y \in \mathcal{M}_{\text{loc}}$. Thus, $X = H \cdot N + Y \in \mathcal{M}_{\text{loc}}$. $\square$

LEMMA 3.8.– Let X be an adapted càdlàg process, $X_0 = 0$. Then

$$\left\{(\Delta X)^-\right\}^* \in \mathcal{A}_{\text{loc}}^+ \quad \Longleftrightarrow \quad \left\{X^-\right\}^* \in \mathcal{A}_{\text{loc}}^+.$$

PROOF.– Since $X^- \leqslant X_-^- + (\Delta X)^-$ and the process X_- is locally bounded, implication $\Rightarrow$ is obvious. Assume that $\left\{X^-\right\}^* \in \mathcal{A}_{\text{loc}}^+$. Put $T_n := \inf\{t\colon X_t > n\} \wedge S_n$, where $\{S_n\}$ is a localizing sequence for $\left\{X^-\right\}^*$. Since $X_- \leqslant n$ on $[\![0, T_n]\!]$, $X - X_- \geqslant -X^- - n$ and $(\Delta X)^- \leqslant X^- + n$ on this stochastic interval. Hence,

$$\left(\left\{(\Delta X)^-\right\}^*\right)^{T_n} \in \mathcal{A}^+,$$

$\left\{ (\Delta X)^- \right\}^* \in \mathscr{A}_{\mathrm{loc}}^+.$ □

Our final result includes theorems 3.19 and 3.20.

THEOREM 3.21.– Let X be a σ-martingale. The following statements are equivalent:

1) $X - X_0$ is a local martingale;

2) there is a special semimartingale Y such that $\Delta X \geqslant \Delta Y$;

3) there is a special semimartingale Y such that $X \geqslant Y$.

PROOF.– Implications (1)⇒(2) and (1)⇒(3) are trivial. To prove the converse implications, let us write the canonical decomposition of the special semimartingale Y in the form $Y = Y_0 + N + B$, where $N \in \mathscr{M}_{\mathrm{loc}}$, $B \in \mathscr{V}$, and B is predictable. Put $M := X - X_0 - N$. It is clear that M is a σ-martingale with $M_0 = 0$. In case (2), $\Delta M \geqslant \Delta B$, where $(\Delta M)^- \leqslant (\Delta B)^- \leqslant \mathrm{Var}\,(B) \in \mathscr{A}_{\mathrm{loc}}^+$. Therefore, by theorem 3.20, M is a local martingale, hence, $X - X_0$ is a local martingale.

Now let (3) hold. Then

$$M \geqslant (Y_0 - X_0) + B.$$

Let $\{S_n\}$ be a localizing sequence for B as an element of $\mathscr{A}_{\mathrm{loc}}$, $T_n := S_n \wedge 0_{\{Y_0 - X_0 < -n\}}$. It is clear that $\{T_n\}$ is a localizing sequence. On the set $\{Y_0 - X_0 < -n\}$, we have $T_n = 0$ and $M^{T_n} = 0$, and, on its complement, $T_n = S_n$ and $M^- \leqslant (Y_0 - X_0)^- + B^- \leqslant n + \mathrm{Var}\,(B)$. It follows that $\left\{ M^- \right\}^* \in \mathscr{A}_{\mathrm{loc}}^+$. By lemma 3.8 and theorem 3.20, M is a local martingale, hence, $X - X_0$ is a local martingale.   □

Appendix

A.1. Theorems on monotone classes

Theorems on monotone classes are a widespread technical tool in stochastic analysis. Here is the scheme of how these theorems are applied. Let a measurable space $(\Omega, \mathscr{F})$ be given. Suppose that we want to check some property for all elements of the σ-algebra $\mathscr{F}$. In other words, if we denote by $\mathscr{D}$ the collection of all subsets of Ω, having the property we are interested in, then our goal is to establish the inclusion

$$\mathscr{D} \supseteq \mathscr{F}. \tag{A.1}$$

Next, we can efficiently check this property on a subclass $\mathscr{C}$ of the σ-algebra $\mathscr{F}$:

$$\mathscr{C} \subseteq \mathscr{D}. \tag{A.2}$$

Moreover, the class $\mathscr{C}$ is wide enough – the smallest σ-algebra it generates is $\mathscr{F}$:

$$\mathscr{F} = \sigma\{\mathscr{C}\}. \tag{A.3}$$

Two theorems given below (on monotone classes and on π-λ-systems) provide sufficient additional conditions on the sets $\mathscr{C}$ and $\mathscr{D}$, that allow us to deduce [A.1] from [A.2] and [A.3]. Two extreme cases of such assumptions are trivial:

– $\mathscr{C}$ is a σ-algebra, no assumptions on $\mathscr{D}$;

– $\mathscr{D}$ is a σ-algebra, no assumptions on $\mathscr{C}$.

The first case is of no interest: it means that $\mathscr{C} = \mathscr{F}$, i.e. we can check directly the property of interest for all elements of the σ-algebra $\mathscr{F}$. The second case is a standard tool. For example, it is used in the reasoning that allows us to conclude that a mapping T from $(\Omega', \mathscr{F}')$ to $(\Omega, \mathscr{F})$ is measurable if $T^{-1}(A) \in \mathscr{F}'$, where A runs over a class $\mathscr{C}$ of subsets of Ω, satisfying [A.3].

DEFINITION A.1.–

1) A collection $\mathscr{D}$ of subsets of Ω is called a *monotone class*, if, for any sequence $A_n \in \mathscr{D}, n = 1, 2, \ldots$, such that $A_n \uparrow A$ or $A_n \downarrow A$, we have $A \in \mathscr{D}$.

2) A collection $\mathscr{C}$ of subsets of Ω is called a π-*system*, if it is stable under finite intersections: if $A, B \in \mathscr{C}$, then $A \cap B \in \mathscr{C}$.

3) A collection $\mathscr{D}$ of subsets of Ω is called a λ-*system*, if (1) $\Omega \in \mathscr{D}$; (2) it follows from $A, B \in \mathscr{D}$ and $A \subseteq B$ that $B \backslash A \in \mathscr{D}$; (3) it follows from $A_n \in \mathscr{D}, n = 1, 2, \ldots$, and $A_n \uparrow A$, that $A \in \mathscr{D}$.

THEOREM A.1 (on monotone classes).– Let $\mathscr{C}$ be an algebra, $\mathscr{D}$ a monotone class, and [A.2] and [A.3] are valid. Then [A.1] holds.

THEOREM A.2 (on π-λ-systems).– Let $\mathscr{C}$ be a π-system, $\mathscr{D}$ a λ-system, and [A.2] and [A.3] hold. Then we have [A.1].

There are several versions of the monotone class theorem for functions. We use the following most widespread version of this theorem in the book.

THEOREM A.3.– Let $\mathscr{H}$ be a linear space of real-valued functions with finite values (respectively real-valued bounded functions) on a set Ω. Assume also that $\mathscr{H}$ contains constant functions and has the following property: for any increasing sequence $\{f_n\}$ of nonnegative functions from $\mathscr{H}$, the function $f = \lim_n f_n$ belongs to $\mathscr{H}$ if it is finite (respectively bounded). Under these assumptions, if $\mathscr{C}$ is a subset of $\mathscr{H}$, which is closed under multiplication, the space $\mathscr{H}$ contains all functions with finite values (respectively bounded functions) that are measurable with respect to the σ-algebra $\sigma(\mathscr{C})$ generated by functions from $\mathscr{C}$.

EXERCISE A.1.– Deduce theorem A.3 from theorem A.2 under an additional assumption that all functions in $\mathscr{C}$ are indicator functions, i.e. take values 0 and 1.

As mentioned above, theorem A.3 contained the additional assumption that $\mathscr{H}$ is closed under the uniform convergence. In fact, this assumption follows from other assumptions of the theorem.

In the application of this result, similarly, we take $\mathscr{H}$ as the class of all functions having the property we are interested in, while $\mathscr{C}$ is a collection of functions for which the property of interest can be directly verified.

A.2. Uniform integrability

Let $(\xi_\alpha)_{\alpha \in A}$ be a family of *integrable* random variables on a probability space $(\Omega, \mathscr{F}, \mathsf{P})$.

DEFINITION A.2.– A family $(\xi_\alpha)_{\alpha \in A}$ of integrable random variables is called *uniformly integrable* if

$$\lim_{c \uparrow +\infty} \sup_{\alpha \in A} \int_{\{|\xi_\alpha| > c\}} |\xi_\alpha| \, d\mathsf{P} = 0.$$

It is clear from the definition that the uniform integrability of a family of random variables is, in fact, a property of their one-dimensional (1D) distributions. Thus, generally speaking, the uniform integrability property may refer to the case where random variables are given on different probability spaces.

A family consisting of a single integrable random variable is uniformly integrable because of the absolute continuity of the Lebesgue integral. Therefore, if each ξ_α satisfies $|\xi_\alpha| \leqslant \eta$, where $\mathsf{E}\eta < \infty$, then a family $(\xi_\alpha)_{\alpha \in A}$ is uniformly integrable.

THEOREM A.4.– A family $(\xi_\alpha)_{\alpha \in A}$ of random variables is uniformly integrable if and only if

$$\sup_{\alpha \in A} \mathsf{E}|\xi_\alpha| < \infty$$

and, for every $\varepsilon > 0$, there is $\delta > 0$ such that $B \in \mathscr{F}$ and $\mathsf{P}(B) < \delta$ imply

$$\int_B |\xi_\alpha| \, d\mathsf{P} < \varepsilon \quad \text{for all } \alpha \in A.$$

Recall that the convex hull of a set X in a linear space consists of elements x that can be represented in the form $x = \sum_{i=1}^{n} \beta_i x_i$, where n is a natural number, $\beta_1, \ldots, \beta_n$ are nonnegative numbers, $\sum_{i=1}^{n} \beta_i = 1$, and $x_1, \ldots, x_n \in X$. The convex hull of X is denoted sometimes by $\operatorname{conv} X$.

THEOREM A.5.–

1) The closure in $L^1(\mathsf{P})$ of the convex hull of a uniformly integrable family of random variables is uniformly integrable.

2) If $(\xi_\alpha)_{\alpha \in A}$ and $(\eta_\beta)_{\beta \in B}$ are uniformly integrable families of random variables, then the family $\{\xi_\alpha + \eta_\beta : \alpha \in A, \ \beta \in B\}$ is uniformly integrable.

THEOREM A.6 (Vallée-Poussin criterion).– Let $(\xi_\alpha)_{\alpha \in A}$ be a family of random variables. The following statements are equivalent:

1) the family $(\xi_\alpha)_{\alpha \in A}$ is uniformly integrable;

2) there is a nonnegative increasing function Φ on $\mathbb{R}_+$ such that

$$\lim_{t \uparrow +\infty} \frac{\Phi(x)}{x} = \infty \qquad \text{and} \qquad \sup_{\alpha \in A} \mathsf{E}\Phi(|\xi_\alpha|) < \infty.$$

Moreover, the function Φ can be taken convex.

The Vallée–Poussin criterion with $\Phi(x) = x^p$ says that every bounded subset of $L^p(\mathsf{P})$, where $p > 1$, is uniformly integrable. This is not true for $p = 1$.

EXERCISE A.2.– Construct a sequence $\{\xi_n\}$ of random variables such that $\sup_n \mathsf{E}|\xi_n| < \infty$, but which is not uniformly integrable.

The following proposition is often used in this book.

PROPOSITION A.1.– Let $(\mathscr{F}_\alpha)_{\alpha \in A}$ be an arbitrary family of sub-σ-algebras of a σ-algebra $\mathscr{F}$, and $\mathsf{E}|\xi| < \infty$. Put $\xi_\alpha = \mathsf{E}(\xi|\mathscr{F}_\alpha)$. Then the family $(\xi_\alpha)_{\alpha \in A}$ is uniformly integrable.

EXERCISE A.3.– Prove proposition A.1 using: (1) definition A.2; or (2) the Vallée–Poussin criterion.

The role of the uniform integrability can be seen from the following theorem.

THEOREM A.7.– Let $\{\xi_n\}$ be a sequence of integrable random variables, and let ξ be a random variable. The following statements are equivalent:

1) the sequence $\{\xi_n\}$ converges to ξ in probability and is uniformly integrable;

2) the random variable ξ is integrable and the sequence $\{\xi_n\}$ converges to ξ in $L^1(\mathsf{P})$, i.e., $\mathsf{E}|\xi_n - \xi| \to 0$ as $n \to \infty$.

In particular, the uniform integrability is sufficient for passing to the limit under the expectation sign in sequences, converging in probability (or a.s.). For sequences of nonnegative random variables, this condition is also necessary.

COROLLARY A.1.– Assume that a sequence $\{\xi_n\}$ of nonnegative integrable random variables converges in probability to a random variable ξ and

$$\lim_{n \to \infty} \mathsf{E}\xi_n = \mathsf{E}\xi < \infty.$$

Then the sequence $\{\xi_n\}$ is uniformly integrable and converges to ξ in $L^1(\mathsf{P})$.

PROOF.– We have

$$E|\xi_n - \xi| = E(\xi_n - \xi) + 2E(\xi - \xi_n)^+.$$

The first term on the right converges to 0 by the assumption. Now note that random variables $(\xi - \xi_n)^+$ are nonnegative, converge to 0 in probability and are majorized by the integrable random variable ξ. Therefore, by the dominated convergence theorem, their expectations converge to 0. Thus, $E|\xi_n - \xi| \to 0$. The uniform integrability of $\{\xi_n\}$ follows from implication (2)⇒(1) in theorem A.7.   □

A.3. Conditional expectation

In this book, we often deal with conditional expectations of random variables which may not be integrable. Working with conditional expectations of such random variables needs some accuracy because they may not be defined or take infinite values. Proposition A.2 allows us to reduce the case where conditional expectations are defined and take only finite values to the case of integrable random variables.

Let a probability space $(\Omega, \mathscr{F}, P)$ and a sub-σ-algebra $\mathscr{G} \subseteq \mathscr{F}$ be given. Recall how the conditional expectation of a random variable ξ with respect to the σ-algebra $\mathscr{G}$ is defined, see [SHI 96] Chapter II, S 7.

First, let $\xi \geqslant 0$. Then the conditional expectation of ξ with respect to $\mathscr{G}$ is a random variable with values in $[0, +\infty]$, denoted by $E(\xi|\mathscr{G})$, such that $E(\xi|\mathscr{G})$ is $\mathscr{G}$-measurable and, for every $B \in \mathscr{G}$,

$$\int_B \xi \, dP = \int_B E(\xi|\mathscr{G}) \, dP. \qquad \text{[A.4]}$$

The existence of conditional expectation follows from the Radon–Nikodým theorem, and it is defined uniquely up to sets of P-measure zero.

The conditional expectation $E(\xi|\mathscr{G})$ of an arbitrary random variable ξ with respect to $\mathscr{G}$ is considered to be defined if, P-a.s.,

$$\min \left(E(\xi^+|\mathscr{G}), E(\xi^-|\mathscr{G}) \right) < \infty,$$

and it is given by the formula

$$E(\xi|\mathscr{G}) := E(\xi^+|\mathscr{G}) - E(\xi^-|\mathscr{G}),$$

where, on the set $\{E(\xi^+|\mathscr{G}) = E(\xi^-|\mathscr{G}) = +\infty\}$ of zero measure, the expression $E(\xi^+|\mathscr{G}) - E(\xi^-|\mathscr{G})$ is defined arbitrarily (keeping $\mathscr{G}$-measurability).

PROPOSITION A.2.– Let ξ be a random variable. The following statements are equivalent:

1) $E(\xi|\mathscr{G})$ is defined and is finite P-a.s.;

2) $E(|\xi| \,|\mathscr{G}) < \infty$ P-a.s.;

3) there exists an increasing sequence of sets $B_1 \subseteq \cdots \subseteq B_n \subseteq \ldots$ such that $B_n \in \mathscr{G}$ for every n, $P(\cup_n B_n) = 1$ and $E|\xi|\mathbb{1}_{B_n} < \infty$ for every n.

If (3) holds, then

$$E(\xi|\mathscr{G})\mathbb{1}_{B_n} = E(\xi\mathbb{1}_{B_n}|\mathscr{G}) \quad \text{P-a.s.} \tag{A.5}$$

Statement (3) means that the equality [A.5] allows us to define $E(\xi|\mathscr{G})$ via the conditional expectations $E(\xi\mathbb{1}_{B_n}|\mathscr{G})$ of *integrable* random variables $\xi\mathbb{1}_{B_n}$.

PROOF.– Implication (1)$\Rightarrow$(2) follows from the definition of the conditional expectation and its additivity for nonnegative random variables, which is quite elementary. If (2) holds, then put

$$B_n = \{E(|\xi| \,|\mathscr{G}) \leqslant n\}.$$

Then (3) holds for this sequence (B_n). In particular, the finiteness of $E|\xi|\mathbb{1}_{B_n}$ follows from the equality [A.4] applied to $|\xi|$ instead of ξ and B_n instead of B.

Let (3) hold. First, assume that $\xi \geqslant 0$. Put

$$\eta := \sum_{n=1}^{\infty} E(\xi\mathbb{1}_{A_n}|\mathscr{G}), \quad A_n := B_n \setminus B_{n-1}, \quad B_0 := \varnothing.$$

Note that

$$\eta\mathbb{1}_{A_n} = E(\xi\mathbb{1}_{A_n}|\mathscr{G}) \quad \text{P-a.s.} \tag{A.6}$$

Since η is $\mathscr{G}$-measurable and, for any $B \in \mathscr{G}$,

$$\int_B \eta \, dP = \sum_{n=1}^{\infty} \int_B E(\xi\mathbb{1}_{A_n}|\mathscr{G}) \, dP = \sum_{n=1}^{\infty} \int_{B \cap A_n} E(\xi\mathbb{1}_{A_n}|\mathscr{G}) \, dP$$

$$= \sum_{n=1}^{\infty} \int_{B \cap A_n} \xi\mathbb{1}_{A_n} \, dP = \int_B \xi \, dP,$$

we have $\eta = \mathsf{E}(\xi|\mathscr{G})$. In view of [A.6], (1) and [A.5] hold.

Now let ξ be an arbitrary random variable. Then (3) holds for ξ^+ and ξ^-. As we have proved, (1) and [A.5] take place for ξ^+ and ξ^-, hence the same is true for ξ. $\square$

A.4. Functions of bounded variation

Here we collect a few results concerning functions of bounded variation that are used in the book. Their proofs can be found in many textbooks on analysis or probability theory. Most of the statements are easy to prove.

Let a function $f\colon \mathbb{R}_+ \to \mathbb{R}$ be given. Define the *variation* of f on an interval $[0, t]$, $t \in \mathbb{R}_+$, by

$$\mathrm{var}_f(t) = \sup \sum_{k=1}^{n} |f(x_k) - f(x_{k-1})|,$$

where the supremum is taken over all n and over all partitions $0 = x_0 < x_1 < \cdots < x_n = t$ of $[0, t]$.

$\mathrm{var}_f(t)$ takes values in $[0, +\infty]$ and is nondecreasing in t.

If f is right-continuous at t and its variation is finite on $[0, t + \varepsilon]$ for some $\varepsilon > 0$, then var_f is right-continuous at t.

If f is right-continuous everywhere on $[0, t)$, then its variation can be computed by the formula

$$\mathrm{var}_f(t) = \lim_{n \to \infty} \sum_{k=1}^{2^n} |f(kt2^{-n}) - f((k-1)t2^{-n})|. \tag{A.7}$$

Let a function $f\colon \mathbb{R}_+ \to \mathbb{R}$ be right-continuous everywhere on $\mathbb{R}_+$, $f(0) = 0$, and let $\mathrm{var}_f(t)$ be finite for all $t \in \mathbb{R}_+$. Put

$$g(t) := \frac{\mathrm{var}_f(t) + f(t)}{2}, \qquad h(t) := \frac{\mathrm{var}_f(t) - f(t)}{2}. \tag{A.8}$$

The functions g and h start from 0, are right-continuous and increasing (in the sense that $s \leqslant t$ implies $g(s) \leqslant g(t)$ and similarly for h), and

$$f = g - h, \qquad \mathrm{var}_f = g + h.$$

The functions g and h are determined by these properties in a unique way. Thus, f can be represented as the difference of two right-continuous functions starting from 0; in particular, there exists a finite limit $\lim_{s \uparrow\uparrow t} f(s)$ at every point $t > 0$.

The converse is also true: the difference of two increasing functions has a finite variation on any finite interval $[0, t]$.

A Lebesgue–Stieltjes measure on $(\mathbb{R}, \mathscr{B}(\mathbb{R}))$ is defined as a measure m on this space such that $m(I) < \infty$ for every bounded interval I. The formula

$$m((0, t]) = f(t) \qquad\qquad\qquad\qquad [\text{A.9}]$$

provides a one-to-one correspondence between Lebesgue–Stieltjes measures vanishing on $(-\infty, 0]$ and increasing right-continuous functions $f \colon \mathbb{R}_+ \to \mathbb{R}$ with $f(0) = 0$. Denote by m_f the Lebesgue–Stieltjes measure corresponding to f. For a measurable function $H \colon \mathbb{R}_+ \to \mathbb{R}$ and $t \in [0, +\infty]$, the Lebesgue–Stieltjes integral

$$\int_0^t H(s)\, df(s)$$

is understood as the Lebesgue integral

$$\int_{(0, t]} H(s)\, m_f(ds);$$

here, if $t = +\infty$, the interval $(0, t]$ is understood as $(0, +\infty)$.

Now let a function $f \colon \mathbb{R}_+ \to \mathbb{R}$ be right-continuous at all points of $\mathbb{R}_+$, $f(0) = 0$, and let $\mathrm{var}_f(t)$ be finite for all $t \in \mathbb{R}_+$. Define functions g and h by relations [A.8], then $m_g + m_h$ is the Lebesgue–Stieltjes measure corresponding to var_f. By Radon–Nkodým theorem, there is a measurable function $H \colon \mathbb{R}_+ \to [0, 1]$ such that, for every $t \in \mathbb{R}_+$,

$$g(t) = \int_0^t H(s)\, d\,\mathrm{var}_f(s), \qquad h(t) = \int_0^t (1 - H(s))\, d\,\mathrm{var}_f(s),$$

then

$$f(t) = \int_0^t (2H(s) - 1)\, d\,\mathrm{var}_f(s).$$

It follows from the previous formula, clearly, that

$$\operatorname{var}_f(t) \leqslant \int_0^t |2H(s) - 1|\, d\operatorname{var}_f(s),$$

hence $(m_g + m_h)(\{s \colon 0 < H(s) < 1\}) = 0$. In other words, there are sets E_+ and E_- from $\mathscr{B}(\mathbb{R}_+)$ such that $E_+ \cap E_- = \varnothing$, $E_+ \cup E_- = \mathbb{R}_+$ i and

$$g(t) = \int_0^t \mathbb{1}_{E_+}(s)\, d\operatorname{var}_f(s), \quad h(t) = \int_0^t \mathbb{1}_{E_-}(s)\, d\operatorname{var}_f(s). \qquad \text{[A.10]}$$

This fact can also be obtained from the Hahn decomposition for a signed measure.

Let a function $f \colon \mathbb{R}_+ \to \mathbb{R}$ be right-continuous at all points of $\mathbb{R}_+$, $f(0) = 0$, $\operatorname{var}_f(t)$ be finite for all $t \in \mathbb{R}_+$, and let $H \colon \mathbb{R}_+ \to \mathbb{R}$ be a measurable function. If

$$\int_0^t |H(s)|\, d\operatorname{var}_f(s) < \infty$$

for some $t \in \mathbb{R}_+$, then we define the integral

$$\int_0^t H(s)\, df(s)$$

by

$$\int_0^t H(s)\, df(s) := \int_0^t H(s)\, dg(s) - \int_0^t H(s)\, dh(s),$$

where the functions g and h are taken from [A.8]. If $\int_0^t |H(s)|\, d\operatorname{var}_f(s) < \infty$ for all $t \in \mathbb{R}_+$, then the function

$$t \rightsquigarrow \int_0^t H(s)\, df(s)$$

starts from zero, is right-continuous, and its variation on $[0, t]$ is equal to

$$\int_0^t |H(s)|\, d\operatorname{var}_f(s)$$

for every $t \in \mathbb{R}_+$.

Let functions $f, g \colon \mathbb{R}_+ \to \mathbb{R}$ be right-continuous at all points of $\mathbb{R}_+$, have finite variation on $[0, t]$ for every $t \in \mathbb{R}_+$, $f(0) = g(0) = 0$. If f and g are increasing functions, we write $df \ll dg$ if the corresponding Lebesgue–Stieltjes measures are absolutely continuous: $m_f \ll m_g$. In the general case, $df \ll dg$ means that $d\operatorname{var}_f \ll d\operatorname{var}_g$.

Bibliographical Notes

This list of references mainly contains textbooks or monographs, where the reader can find an alternative development of the results, missing proofs etc. It is by no means exhaustive; a number of excellent books on the subject are not included. No attempt to specify the original sources is done. The lack of a citation does not imply any claim to priority on my part.

Chapter 1

General references: [DEL 72, DEL 78, ELL 82, HE 92, JAC 03, LIP 89, MED 07, MÉT 82, YEH 95].

Section 1.1. Concerning continuity of the stochastic basis generated by Lévy processes or, more generally, by a process with independent increments (Remark 1.1), see [KRU 10].

Section 1.3. For the proof of Theorem 1.7 see, e.g., [DEL 72] or [DEL 78]. Proposition 1.12 is taken from [HE 92].

Section 1.4. The proof of Theorem 1.12 can be found in [DEL 78].

Chapter 2

General references: [BIC 02, DEL 72, DEL 82, ELL 82, HE 92, JAC 79, JAC 03, LIP 89, MED 07, MÉT 82, MEY 76, PRO 05, YEH 95].

Section 2.1 The proofs of theorems 2.1, 2.2, 2.3, 2.5 can be found in many sources, in particular, in [DEL 82, ELL 82, HE 92, KAL 02, MED 07, YEH 95].

Section 2.7. The property [2.52] of the quadratic variation is proved, e.g., in [DEL 82, HE 92, JAC 03, MED 07, MEY 76, PRO 05, YEH 95].

Section 2.8. For the proofs of the Burkholder–Davis–Gundy inequality see, e.g., [DEL 82, HE 92, LIP 89].

Chapter 3

General references: [BIC 02, DEL 82, ELL 82, HE 92, JAC 79, JAC 03, LIP 89, MED 07, MÉT 82, MEY 76, PRO 05, YEH 95].

Section 3.2. For the proofs of Itô's formula see, e.g., [DEL 82, HE 92, MED 07, MEY 76, PRO 05, YEH 95].

Section 3.5. The form of the Ansel–Stricker theorem presented in theorem 3.21 seems to be new.

Appendix

Section A.1. Proofs of Theorems A.1–A.3 can be found, e.g., in [DEL 78, SHA 88, SHI 96, KAL 02].

Section A.2. For more details, see, e.g., [DEL 78, SHI 96, KAL 02].

Section A.4. See, e.g. [KAL 02].

Bibliography

[BIC 02] BICHTELER K., *Stochastic Integration with Jumps*, Encyclopedia of Mathematics and its Applications, Cambridge University Press, Cambridge, 2002.

[DEL 72] DELLACHERIE C., *Capacités et processus stochastiques*, Series of Modern Survey in Mathematics, Springer-Verlag, Berlin, vol. 67, 1972.

[DEL 78] DELLACHERIE C., MEYER P.-A., *Probabilities and Potential*, North-Holland Mathematics Studies, North-Holland Publishing Co., Amsterdam, New York, vol. 29, 1978.

[DEL 82] DELLACHERIE C., MEYER P.-A., *Probabilities and Potential. B*, North-Holland Mathematics Studies, North-Holland Publishing Co., Amsterdam, vol. 72, 1982.

[DEL 06] DELBAEN F., SCHACHERMAYER W., *The Mathematics of Arbitrage*, Springer-Verlag, Berlin, 2006.

[ELL 82] ELLIOTT R.J., *Stochastic Calculus and Applications*, Applications of Mathematics, Springer-Verlag, New York, vol. 18, 1982.

[HE 92] HE S.W., WANG J.G., YAN J.A., *Semimartingale Theory and Stochastic Calculus*, Science Press, Beijing, and CRC Press, Boca Raton, FL, 1992.

[JAC 79] JACOD J., *Calcul stochastique et problèmes de martingales*, Lecture Notes in Mathematics, Springer, Berlin, vol. 714, 1979.

[JAC 03] JACOD J., SHIRYAEV A.N., *Limit Theorems for Stochastic Processes*, Comprehensive Studies in Mathematics, Springer-Verlag, Berlin, 2nd edition, vol. 288, 2003.

[KAL 02] KALLENBERG O., *Foundations of Modern Probability*, Probability and its Applications, Springer-Verlag, New York, 2nd edition, 2002.

[KRU 10] KRUGLOV V.M., "On the continuity of natural filtrations of processes with independent increments", *Theory of Probability and its Applications*, vol. 54, no. 4, pp. 693–699, 2010.

[LIP 89] LIPTSER R.S., SHIRYAEV A.N., *Theory of Martingales*, Mathematics and its Applications, Kluwer Academic Publishers Group, Dordrecht, vol. 49, 1989.

[MED 07] MEDVEGYEV P., *Stochastic Integration Theory*, Oxford Graduate Texts in Mathematics, Oxford University Press, Oxford, vol. 14, 2007.

[MÉT 82] MÉTIVIER M., *Semimartingales*, de Gruyter Studies in Mathematics, Walter de Gruyter & Co., Berlin-New York, vol. 2, 1982.

[MEY 76] MEYER P.A., *Un cours sur les intégrales stochastiques*, Séminaire de Probabilités X, pp. 245–400, Lecture Notes in Mathematics, vol. 511, Springer, Berlin, 1976.

[PRO 05] PROTTER P.E., *Stochastic Integration and Differential Equations*, Stochastic Modelling and Applied Probability, Springer-Verlag, Berlin, vol. 21, 2005.

[SHA 88] SHARPE M., *General Theory of Markov Processes*, Pure and Applied Mathematics, Academic Press, Inc., Boston, MA, vol. 133, 1988.

[SHI 96] SHIRYAEV A.N., *Probability*, Graduate Texts in Mathematics, Springer-Verlag, New York, 2nd edition, vol. 95, 1996.

[SHI 99] SHIRYAEV A.N., *Essentials of Stochastic Finance*, Advanced Series on Statistical Science & Applied Probability, World Scientific Publishing Co. Inc., River Edge, NJ, vol. 3, 1999.

[SHI 02] SHIRYAEV A.N., CHERNY A.S., "Vector stochastic integrals and the fundamental theorems of asset pricing", *Proceedings of the Steklov Institute of Mathematics*, vol. 237, pp. 6–49, 2002.

[YEH 95] YEH J., *Martingales and Stochastic Analysis*, Series on Multivariate Analysis, World Scientific Publishing Co., Inc., River Edge, NJ, vol. 1, 1995.

Index

Printed in the United States
By Bookmasters